Introduction to Computational Origami

Ryuhei Uehara

Introduction to Computational Origami

The World of New Computational Geometry

Ryuhei Uehara
JAIST
Ishikawa, Japan

ISBN 978-981-15-4472-9 ISBN 978-981-15-4470-5 (eBook)
https://doi.org/10.1007/978-981-15-4470-5

This Springer imprint is published by the registered company Springer Nature Singapore Pte Ltd.
The registered company address is: 152 Beach Road, #21-01/04 Gateway East, Singapore 189721, Singapore

Preface

In Japan, origami is often regarded as a kind of child's play. Considered in this way, origami is not mentioned widely in Japan as a part of science. In Japanese, the word "*ori*" means "folding" and *gami* means "paper". That is another reason why the word "origami" has limited meaning in Japan. "Origami" is now commonly used in English, and this word has been gaining a broader meaning than the original Japanese word. One can find many variants of origami books, which are wider ranging than Japanese ones, on the shelves of the origami section at bookstores in other parts of the world. The English word "origami" represents a wider range of paper crafts than in Japan, and it seems that this range is still expanding.

From the industrial point of view, the application of the concept of origami in a broad sense is definitely spreading, however. For example, "Miura map folding", invented by Koryo Miura, is a famous origami folding that applied to the foldable structure of the huge solar panels of a space station. Now it is used for more general architecture by extending the folding pattern to more general foldable patterns. There is "the Yoshimura pattern" which is used for some drink cans and that is useful for reducing the material while maintaining their strength. For stent grafts, which are kinds of artificial blood vessels, those made by shape memory alloys with an origami technique have been developed. In some areas, not noticed immediately, such as research on folding robots, bodies of cars, and DNA folding, investigation of folding is attracting much more attention than previously. Just as the Japanese word "origami" has become understood in English, child's play has also become science with wide applications.

With the development of these examples of "origami science", since 1989 an international conference on origami science, mathematics, and education has been held almost every 4 years. Specifically, it was held in Ferrara, Italy (December 1989), Shiga, Japan (November 1994), Monterey, USA (March 2001), Pasadena, USA (September 2006), Singapore (July 2010), Tokyo, Japan (August 2014), and Oxford, UK (September 2018). Starting with the third conference, they have been referred to as 3OSME, 4OSME, 5OSME, 6OSME, and 7OSME, respectively (OSME stands for *Origami in Science, Mathematics, and Education*). These conferences focus on the three major topics of science, mathematics, and education. In recent years, origami science has had a strong presence, and it is especially impressive that research using computers has been quite active.

Following this trend, interest in origami in the society of computer science has also emerged. In recent years, occasions to hear the term "computational origami" have increased. Computational origami is a field of computer science, especially computational geometry. It is an active research area, and many talks on the subject have been given at recent international conferences in computational geometry.

This book is an introduction to computational origami. In the first place, computational origami is a very new research field. In the 1990s, research began on the relationship between a polygon and a polyhedron folded from it, and that research area was named "computational origami". Considering that whole problems with "folding" and "unfolding" as the basic operations comprise the research subjects, the range of this area is extraordinarily wide and has many applications. By contrast, the population of researchers is still small, so many incomplete fields remain to be investigated. Some researchers say that any research topic will lead to a paper. Based on the current situation, this book aims to facilitate an understanding of the forefront of computational origami. It can be used as a textbook for universities and graduate schools and is devised to be interesting to engineers who are considering the use of origami engineering and to general readers who are interested in the application of computer science.

Prerequisite knowledge is not necessarily required, and in fact, there are many parts that require only elementary mathematics at the high school level. Rather, like relationships between solids and their unfoldings, there are many themes that require recalling what was learned at elementary school. Actually, something like what one learned at elementary school is deeper than might be imagined, and that point, first of all, may be surprising. For example, in the case of unfoldings, sometimes there are currently no solutions other than finding all the cases by using computers finally. However, everyone can understand the solutions themselves for such problems. Such a gap is also one of the interesting points in computational origami. For people who are interested in this kind of puzzle-like unfolding, it may be quite enjoyable to just look at and confirm the results. In any case, the reader will be able to overview the kinds of problems the modern computational origami researchers approach. Students trying to start researching computational origami by themselves can find a suitable theme to begin on. In the field of computational origami, there are many examples in which students (graduate students, undergraduate students, sometimes high school students!) have played an important role in finding solutions. It will be an unexpected pleasure if this book helps encourage the activities of such ambitious students.

This book is largely based on my Japanese book published by Kindai-Kagaku-sha in 2018. Most but not all the contents are from journal papers and conference papers written with my coauthors, including Zachary Abel, Yoshiaki Araki, Brand Ballinger, Therese Biedl, Jean Cardinal, Timothy M. Chan, Erik D. Demaine, Martin L. Demaine, David Eppstein, Jeff Erickson, Adam Hesterberg, Takashi Horiyama, Jinfeng Huang, Shinji Imahori, Hiro Ito, Tsuyoshi Ito, Masashi Kiyomi, Irina Kostitsyna, Stefan Langerman, Anna Lubiw, Jayson Lynch, Hiroaki Matsui, Jun Mitani, Koichi Mizunashi, Yuta Nakane, Paul Nijjar, Günter Rote, Toshihiro Shirakawa, Jack Snoeyink, Takuya Umesato, Takeaki Uno, Yushi Uno, Toshiki Saitoh, Ming-wei Wang, Dawei Xu, and Tomoo Yokoyama. The origami font on the cover of this book is called *Impossible Folding Font* and is by Erik D. Demaine, Martin L. Demaine, Tomoko Taniguchi, and Ryuhei Uehara, 2019. See https://erikdemaine.org/fonts/impossible/ for further details.

I greatly appreciate all of these individuals.

Kanazawa, Japan
March 2020

Ryuhei Uehara

About This Book

This is a book on computer science, especially "computational origami" which is a trend of recent computational geometry. There are few opportunities for origami to be expressed as a computer science research even in Japan. Meanwhile, from a broad perspective, the operation of "folding" some material is everywhere around the world on a daily basis. Everyone "folds" or "bends" something in everyday life. The viewpoint of considering this familiar thing as mathematical research subject may be novel.

In the first place, computer science is the science that deals with the process of computation, that is, it is a research area thinking of a good method of computation, or algorithms. For example, when programmers write some programs to obtain the same results, it is a common practice that there is a world of difference between a good programmer and a poor programmer in execution time and amount of memory. That is the algorithm, or the method of computation, which determines the efficiency of the computation.

Origami, which stands for Japanese "ori (folding)" and "kami (paper)", is now an international word that is accepted in English as it is. The basic operation in origami is, of course, "folding". "Computational origami" is an "Origami Science" from the viewpoint of computer science, and it can be considered as a research on algorithms with folding as a basic operation. For example, in order to fold the desired shape and/or pattern, what kind of folds and in what order should be applied? Also, when folding origami, if you pile sheets of paper and fold them together, you will be able to reduce the number of folds. Nonetheless, it does not seem to be a simple matter to make the mountains and valleys of crease lines exactly as you thought when folding them. In addition, if you fold too many piles, you will have a margin of error by their thickness. Come to think of "folding" as a basic operation, there are various problems with it. This book focuses on origami from the viewpoint of computer science and introduces from the basic theorems to the latest results.

Contents of This Book

In this book, we pick up two major topics from a number of themes related to "computational origami" and summarize the vivid research results recently obtained.

The first topic is "development", which is also called "net", of a solid. Speaking of nets of solids, most readers will be reminded that you learned the notion when you were elementary school students. However, it was just a prelude to the fertile field. There are numerous unresolved problems hidden there. For example, in elementary schools in Japan, they learn that there are 11 nets in a cube. Actually, for the regular octahedron, there are 11 nets, and there are only two types of regular tetrahedron. Some people may know these facts. However, no one remembers whether you learned the number of nets of a regular icosahedron and a regular dodecahedron. In fact, there are 43380 types of nets in both of them. To be more precise, it has been known for quite a long time that there are 43380 different ways of developments. However, it was not confirmed that all of these developed shapes had truly no overlapping until 2011 when this was unveiled. This is considerably surprising! We know very little about the nets of solids. Also, note that the number of such "nets" is considered only for the cases that the cut is made along the edges of the solids. In this book, we will forget this constraint that we should cut along only edges. Even for a cube and a regular tetrahedron, it turns out that there are infinite nets.

For these infinite nets, a beautiful mathematical characterization for a regular tetrahedron using the notion of tiling is known. However, we do not know much about the relationship between other solids and their nets. Although it is a mathematical result that there are 43380 ways of developing a regular icosahedron and a regular dodecahedron in both cases, it cannot be confirmed without the help of a computer that each of them does not actually overlap when it is unfolded. In this research area, the mathematical characterization and the development of efficient algorithms on computer are closely connected with each other. This is exactly the aspect of computer science.

The second topic is "folding models and their computational complexity". From the viewpoint of computer science, "computational origami" is a considerably new research field, and there is no established computation model that is the common ground for discussion. In accordance with the problem handling on that occasion, we have devised an appropriate model and conducted research on that basis while examining validity.

For example, one-dimensional origami, which is folded along equally spaced creases, would be considered as the simplest origami model. In other words, it is a model that there are creases at regular intervals in an elongated ribbon-shaped origami. Even in such a simple model, many open problems are still left. For example, given that you fold along equally spaced n creases on the one-dimensional origami into a minimum unit length, how many essentially different ways are there?

This is a problem called a stamp folding problem, and any specific function is not known.

Extending the simple model, we can obtain many different general models; for example, the regular two-dimensional origami, folding crease lines not equally spaced, and/or crease lines are inclined diagonally. We can consider a number of problems on these realistic models, but we do not know anything exactly. Here, I did not mention paper thickness at all; however, for example, given a method of folding thick material well, that is a widely applied topic, only a few basic things are understood now.

When such one basic folding model is fixed, the efficiency of folding is exactly the research of computer science. If it is difficult, it is intractable in terms of computational complexity, and conversely, if there exists an efficient way of folding, it will be proposed as a good algorithm. Computer science and origami are quite compatible.

After introducing the above two topics, I will present the latest research topics as advanced problems. With the latest research results, you will see what we know and what problems are still open.

In the last chapter, I give commentaries on all exercises. Also, I post some origami-related puzzles in places. I hope readers enjoy these things at the same time.

Assumed Reader in This Book

This book introduces the latest research results of origami science from the viewpoint of computer science. There are some places where knowledge about computational complexity and/or algorithm theory is assumed in some topics related to theoretical computer science. However, we do not premise any particular prerequisite knowledge in most of the contents. If you read the basics in the first half of this book, college students and motivated high school students will understand a lot of parts. The second half introduces the latest research results and unresolved problems related to them. Especially, if you are a student in information science department or a graduate student, you will be able to find research themes here and there.

Also, aside from finding out how to find it, if you actually fold the various diagrams contained in this book and try to create a resulting solid, you will be able to enjoy the wonders of mathematics. This book includes many strange nets that are contrary to intuition and no one has ever seen before, it is worth a try.

The research field of computational origami has still short history, and many of the topics have not been studied much in the conventional framework. With youthful and flexible ideas, college students and even high school students can sometimes produce amazing results in such fields. Hoping for it, this book introduces a number of exercises and unresolved problems. There may be problems that can be solved quickly with sharp insights by a flexible smart reader.

Constructs of This Book

This book is largely divided into four parts.

Part I is the basic knowledge of the geometry of the nets. Speaking of nets, it may be the last thing at elementary school, but there are various interesting properties and mathematical features. Let us consider the properties of these nets again.

In Part II, we will learn about the various outcomes of the nets of the solids. In particular, let us learn from a variety of perspectives about the solid Q that can be folded from some polygon P. If you are interested in figures, especially nets of solids, it would be better to read this part. If you have not thought of the nets since you learned in elementary schools, I bet that this part is full of big surprises.

Part III is the topic of complexity and algorithms on a simple origami model. This is a bit-independent topic from the above nets and especially recommended to readers who are interested in computer scientific aspects of computational origami.

Part IV is a collection of related topics in random order, but many of them are related to the previous part. Let us look at further details below.

In Chap. 7, we consider the problem that asks if we can fold a (flat) petal shape into a pyramid. If you understand the fundamental part of the nets of Part I, you can read this chapter independently. Topics on computational geometry such as Voronoi diagram, which appears to be irrelevant to origami, are quite cleverly used in solving the folding problem. It will be enjoyable for readers interested in computational geometry.

In Chap. 8, the nets are considered when they are obtained by cutting along lines without branch in the middle. The relationship between the solid and the net dealt with here is a model that can be realized unfolding/folding; for example, a zipper, and we can actually see such a craft as well. This chapter is also relatively highly independent.

Chapter 9 deals with a packing problem of a net of a cube by a set of nets of a (smaller) cube, which is almost like a puzzle. Anyone who is familiar with puzzles so-called "polyomino" will enjoy this topic.

Chapter 10 is a developed topic of Part II. It deals with nets of a wide variety of solids such as the Johnson-Zalgaller polyhedra. In particular, it is fun for people who like not only regular polyhedra but also similar polyhedra. In addition, we introduce various algorithmic techniques for dealing with these nets of solids, including quite recent techniques. For example, if you are interested in techniques that efficiently deal with astronomically many nets, you will get a variety of useful information.

Chapter 11 is an advanced topic of Part III. Almost all "computational origami" in this book can be handled in the world of natural numbers. However, the sheet of paper is a two-dimensional plane, which is originally supposed to be in the world of real numbers. When you carefully consider this "difference", you get to the gap in a very intrinsic computation model. This gap is related to very deep themes of theoretical computer science, such as "Gödel's incompleteness theorem", "countable and uncountable sets", and "computable and general functions". The key term to this discussion is the "diagonalization". If you are interested in such a topic, you will enjoy this chapter deliberately.

Support Website

In this book, there are plenty of nets full of surprises, which you cannot easily understand without cutting them out and folding them actually. Some of them will be improved the visibility if they are colored. I maintain a supplemental website so that you can actually enjoy cutting out such nets. I add the latest information and supplemental information at the URL below.

http://www.jaist.ac.jp/uehara/books/NewGeometry/

In addition, the various polyhedral data used in this book are the data published by Prof. Jun Mitani of Tsukuba University on his website:

http://mitani.cs.tsukuba.ac.jp/polyhedron/

Contents

Part IV Advanced Problems

Part I
Introduction to Unfolding

Chapter 1
Unfolding

Abstract Part I and Part II introduce various results of the unfoldings that are introduced. First of all, what is an "unfolding"? When you cut the surface of a solid and spread it over a plane, you may have a "net" of the solid by this unfolding. To obtain a proper net, it is required to have two important properties; it should be one piece, and it has no overlapping. These simple properties are sometimes quite difficult to have.

1.1 Unfolding and Edge-Unfolding

First of all, what is a "unfolding"? Many people will imagine something like polygons in Fig. 1.1. Actually, when we were in elementary school, we learned that "there are eleven different types of nets by cutting and unfolding a cube". Then, have you considered this reverse? Is it possible to obtain only the cubes by folding these 11 polygons?

The polygon in Fig. 1.2 is one of the 11 nets of the cube in Fig. 1.1. Folding this polygon along the lines shown in the figure, surprisingly, you obtain a sharp tetrahedron. All four sides are triangles. In fact, from this polygon called Latin Cross, 23 types of different convex polyhedrons can be folded with 85 kinds of different ways of folding [DO07]. In addition to the cube and this pointed tetrahedron, there are 21 more types that can be folded. This is a surprise! In this book, we consider that these relationships are "polyhedron" and "net" which can be obtained by unfolding the polyhedron. The point is that "you can cut in the face of the polyhedron". With this premise, it is immediately apparent that there are infinitely many nets of a cube (Fig. 1.3).

We here introduce some notions dealt with in this book. A *polygon* is a shape that consists of finite line segments. We usually consider *simple polygon*; that is, the set of line segments makes a closed cycle, and no pair of line segments intersects with each other. A polygon has inside and outside by the closed cycle, and it has no hole inside. A *polyhedron* is a solid that consists of finite faces, and each face is a polygon. A polyhedron has inside and outside by the set of faces. Here, we introduce key notions in this book:

R. Uehara, *Introduction to Computational Origami*,
https://doi.org/10.1007/978-981-15-4470-5_1

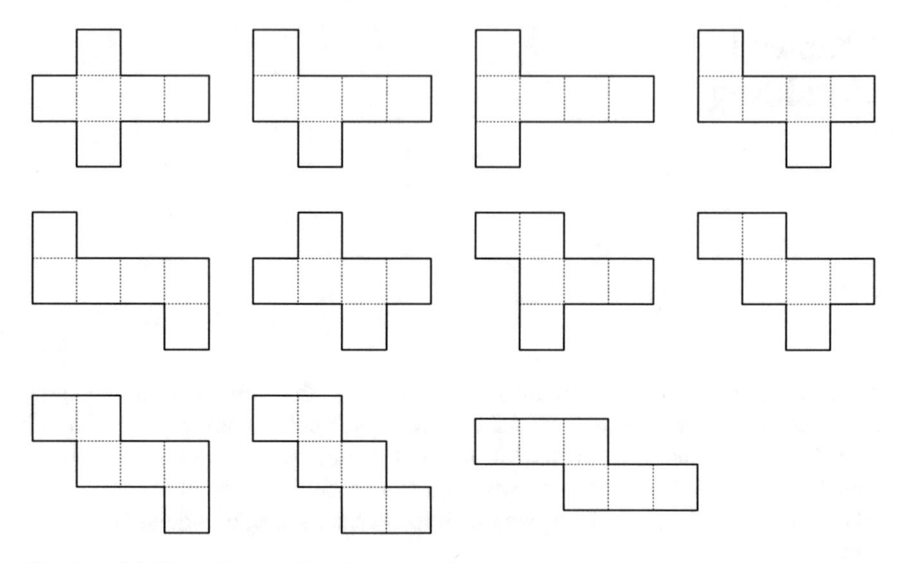

Fig. 1.1 Unfolding of a cube along the edges: 11 nets

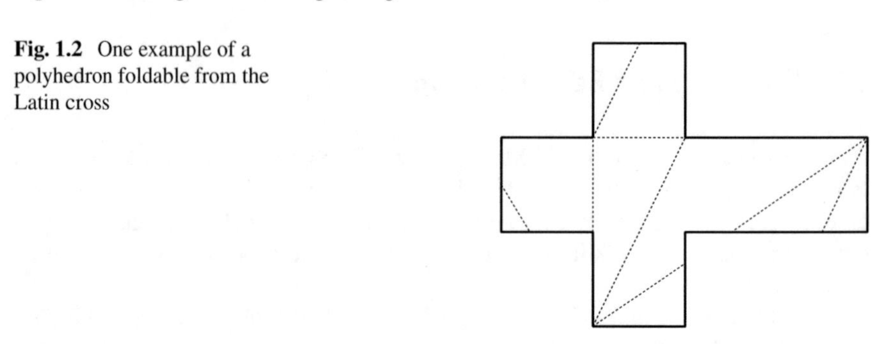

Fig. 1.2 One example of a polyhedron foldable from the Latin cross

Definition 1.1.1 With respect to the polyhedron Q, we cut incisions on edges and faces, and unfold it. Then the obtained polygon P is called a *net* of Q if P is *connected*, and *there is no overlap*.

We here note that we sometimes use the other terms *unfolding* and *development* in a similar meaning. When we "unfold" a polyhedron Q by some cut lines, we have not yet known that if we can obtain a connected simple polygon with no overlap. By unfolding Q to flat, we obtain some "development" P. If P is connected (or one piece) simple polygon with no overlap, then we can say that P is a net of Q. In computational origami area, these terms are not yet unified. In this book, I tried to unify these technical terms.[1] Note that a word "development" has bit a different meaning when we use it as a "developable surface" in a similar context. Therefore, we usually use "unfold" instead of "develop" in this book.

[1]I thank Professor Joseph O'Rourke for his kind suggestions and discussions about these terms.

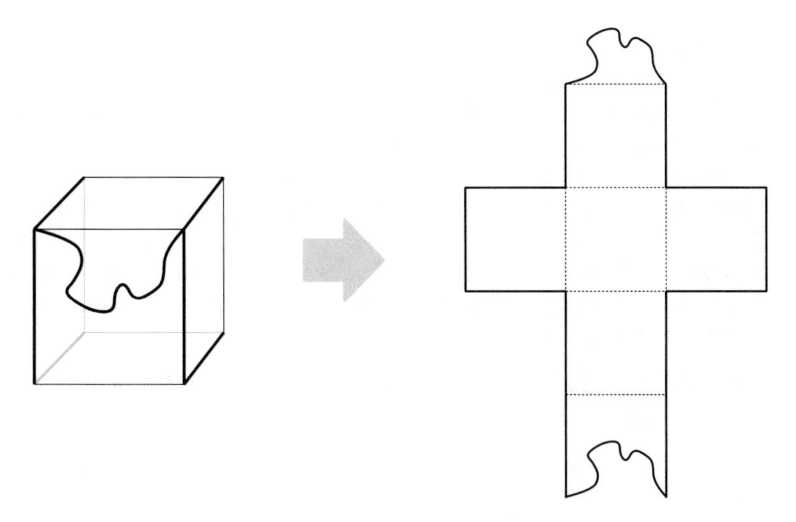

Fig. 1.3 One example of infinitely many nets of a cube

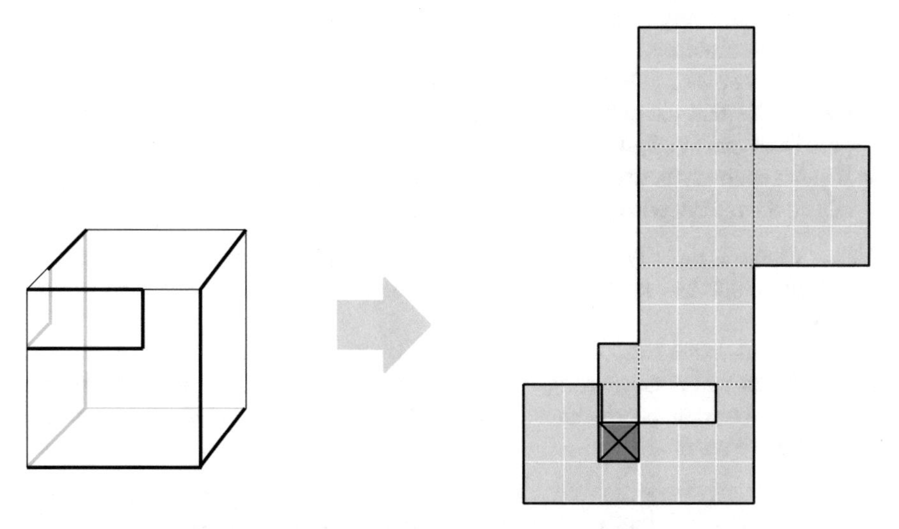

Fig. 1.4 An unfolding of a cube with overlapping: unfolding after cutting along the lines shown in the left figure, it overlaps at the squares marked by $\times$ in the right figure

In this book, we assume that a face of a solid is flat, and each edge of the face is a straight line. (Of course there are curved surfaces, curved solids, and nets, but within the scope of this book, it is a very challenging problem.) The points to note here are that it is connected and it does not overlap. The former point is intuitively obvious. If you can cut into any number of pieces, it is easy to unfold any polyhedron into flat pieces without overlap. The latter point may seem natural and trivial, but there is a quite counterintuitive point here. For example, even a cube can overlap when opening properly a notch and unfolding it. A simple example is shown in Fig. 1.4.

(It may not be that simple. If you are not convinced, it is definitely better to try and see it. It is easier to make each side a square of 3×3, so I draw the reference lines in the development on the right side of the figure.) Connectivity of polygons can be checked with comparative ease, but it is generally difficult to determine overlapping of such surfaces. It is one of the key points in the research of unfoldings.

In this book, when we say "unfold", it is allowed that we can cut in the faces. Usually, unfoldings that we normally think are cut along edges. Such a kind of unfoldings is called *edge-unfolding* in particular. For example, 11 nets shown in Fig. 1.1 are all of the edge-unfoldings of a cube. There is the biggest open problem of computational origami concerning edge-unfolding:

Problem 1.1.1 (*Edge-unfolding conjecture*) Any convex polyhedron has always a net by edge-unfolding.

Currently, we know very little about this conjecture. Indirectly, for example, the following facts are known (details can be found in [DO07]).

- There are counterexamples if they are allowed to be concave. Some type of polyhedron (intuitively a solid with many sharp parts like a sea urchin) overlaps for any edge-unfolding.
- It is always possible to unfold by general unfolding, not edge-unfolding. (A theorem states that the unfolding does not overlap when you cut along the shortest path from one fixed general point to every vertex of the polyhedron.)
- If you randomly generate a convex polyhedron and randomly unfold along the edges, it overlaps with probability 1 experimentally.

None of them is directly related to the open problem, and there is evidence that the conjecture might be established and it will not be established, so there is no wonder in either way.

At present, some results are known about cases where restrictions are imposed on polyhedra and nets. Starting from this limited place, we will try to clarify the relationship between polyhedra and nets step by step. In this chapter, we will present some of these results.

Difficulty of Developments of a Convex Polyhedron (Puzzle Question)

Some readers may think that it is a simple matter to unfold a convex polyhedron. For example, such a reader would like to consider "an unfolding of a cube made of six squares" that is *not* shown in Fig. 1.1. At first, you may think "Is there such a thing?" However, if you notice the possibility that "the size of a square is not necessarily one of a unit square", there is a surprising solution. It is also known that there is an infinite solution even if "the six squares are all the same size". As far as I know, this is the puzzle originally invented by a puzzle designer Masaka Iwai. The answer is somewhere in this book, but I would like you to take the time to consider it first.

Reference

[DO07] E.D. Demaine, J. O'Rourke, *Geometric Folding Algorithms: Linkages, Origami, Polyhedra* (Cambridge University Press, Cambridge, 2007)

Chapter 2
Basic Knowledge of Unfolding

Abstract First of all, as a basic knowledge, we learn that unfoldings can be characterized by spanning trees. We point out the relation with graph theory. Next we show the theorem by Akiyama and Nara. This is a beautiful mathematical theorem which shows the relationship between the unfolding of a tetrahedron and tiling.

2.1 Basic Properties of Unfolding

In graph theory, *tree* refers to a connected graph without cycles. In any given connected graph, we call the tree that connects all vertices of the graph *spanning tree*. There is a close relationship between the notion of the spanning tree and any unfolding of polyhedron.

First, we consider the general unfolding of a convex polyhedron Q. Let C be the set of cut lines of the convex polyhedron Q, and P be the polygon obtained by unfolding Q. Then we have the following.

Theorem 2.1.1 *The set of cut lines C on Q forms a spanning tree connecting all the vertices on Q. The outer circumference of the polygon P is twice the total length of line segments in C.*

Proof Considering each vertex of the convex polyhedron Q, the sum of the angles of the polygons gathered around this vertex is less than 360° because the area around this vertex is not flat. Therefore, in order to unfold it on a flat plane, this vertex must be necessarily cut. Thus C must pass through all vertices. If C has a cycle, P is divided into the inside and outside of the cycle, so it cannot be a net. Therefore, C has no cycle. Thus, it is sufficient to show that C is connected. Assume that C contains two (or more) connected components, say, C_1 and C_2. Then there should be an annular sequence of faces separating C_1 and C_2 on Q as shown in Fig. 2.1. Since this can be opened flat, either C_1 or C_2 must be on the plane. In that case, cutting on that side is unnecessary, which is a contradiction. Thus C is a single spanning tree.

When the line segments in C are cut, each line segment in C is split into two line segments on the boundary of P. From this fact, it is obvious that the outer circumference of P is twice the total length of the line segments in C. □

R. Uehara, *Introduction to Computational Origami*,
https://doi.org/10.1007/978-981-15-4470-5_2

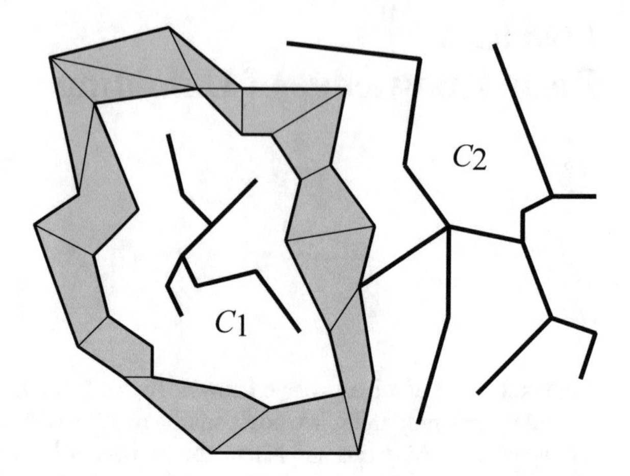

Fig. 2.1 Case that cut lines are divided into two or more sets. If there is a sequence of faces (gray in the figure) that separates C_1 and C_2, this sequence will be unfolded onto the plane by itself, which is a contradiction

Theorem 2.1.1 is also valid for edge-unfolding. That is, the set C of cut lines giving an edge-unfolding of the convex polyhedron Q is a spanning tree of the graph induced by the vertices and edges of Q. According to a result of the graph theory, any tree of n vertices always has $n - 1$ edges. Therefore, the following corollary is obtained.

Corollary 2.1.2 *Let Q be any polyhedron that consists of n vertices and edges of unit length. Then the circumference of any net of Q is of length $2n - 2$.*

For example, the circumferences of the nets in Fig. 1.1 are all $2 \times 8 - 2 = 14$. That is, in the regular polyhedra, cutting lines cannot be saved as long as they are concerned with edge-unfoldings. For example, it is known that a regular tetrahedron has two edge-unfoldings, but both have circumference 6.

However, if it is not concerned with edge-unfolding, it can be developed with shorter cut lines.

Exercise 2.1.1 *In the (general) unfolding of a regular tetrahedron, find the one with the shortest line length in total to cut.*

The answer to this exercise, "unfolding of the regular tetrahedron with the shortest cutting lines", is quite interesting in shape. Can you find by yourself?

The following simple theorem is sometimes very useful in the discussion on the unfolding.

Theorem 2.1.3 *Let P be a net of a convex polyhedron Q. Then the vertices of Q are on the boundary of P.*

Proof Let p be any vertex of the convex polyhedron Q. Then the sum of the sheet of paper around p is less than 360°. Therefore, we cannot make it flat without cutting this point p. That is, the cut lines to unfold Q should pass through p. As a result, p is a point on the boundary of P. We note that p can be two or more points on the boundary of P when a cut line contains p inside, or the set of cut lines makes a branch at p. □

2.2 The Number of Edge-Unfoldings

Now, we consider the nets of regular polyhedra as typical polyhedra. A *regular polyhedron* is a polyhedron in which all faces are congruent regular polygons, and the same number of regular polygons is gathered around each vertex. Specifically, there are five types of regular polyhedra: regular tetrahedron, cube, regular octahedron, regular dodecahedron, and regular icosahedron.

Exercise 2.2.1 *Prove that there are five regular polyhedrons.*

These five types of polyhedra are also called *Plato solids*. Since ancient times, they have been well studied in all aspects. What about their nets, then? Considering the general nets, as shown in Fig. 1.3, there are uncountably many nets. On the other hand, considering the edge-unfoldings, the numbers are finite. As already introduced, there are two edge-unfoldings of a regular tetrahedron, and there are 11 edge-unfoldings of a cube. Then how about a regular octahedron? Actually, it has 11 edge-unfoldings. Surprisingly, a regular icosahedron has 43380 kinds of edge-unfoldings, and a regular dodecahedron also has 43380 kinds of edge-unfoldings.

It can be confirmed that a regular tetrahedron has two edge-unfoldings, and both a cube and a regular octahedron have 11 edge-unfoldings. However, it is quite surprising to hear that each of a regular icosahedron and a regular dodecahedron has 43380 edge-unfoldings. Originally, who unfolded and confirmed that they did not have overlap? Actually, the number of ways of unfoldings for each regular polyhedron has been known since a long time ago. Specifically, we can count the number of the spanning trees of the corresponding graphs based on Theorem 2.1.1, which is not so new. However, we had not known whether each edge-unfolding really has no overlap until quite recently. In 2011, it was checked by computer for the first time that whether each of all edge-unfoldings overlaps or not, and it was confirmed that all the ways of unfoldings indeed produce the net, or the edge-unfoldings without overlap [HS11]. Intuitively, this may seem to have merely confirmed trivial results, but that is not the case. For example, a polyhedron with the same structure like a familiar soccer ball is called a *truncated icosahedron* (see Fig. 4.3), but it is known that it can overlap when cutting it along certain edges. Moreover, it is quite often, contrary to intuition. For the truncated icosahedron, there are 3127432220939473920 different ways of edge-unfolding in total [HS13]. Since this is a very large number, it is still unknown that how many edge-unfoldings do not have overlap. That is, the number of nets of a soccer ball by edge-unfolding is not known.

Now we turn to the numbers of edge-unfoldings of the regular polyhedra. Did you notice that the numbers of edge-unfoldings of a cube and a regular octahedron are the same 11? You may think this is a coincidence. Then, what do you think about the fact that both the regular dodecahedron and the regular icosahedron have the same number 43380 of edge-unfoldings? Of course, this is not a coincidence. Because this is an interesting property of the unfoldings, let us examine the fact more thoroughly.

First, we introduce the notion of *dual* in polyhedra. The *dual* of a polyhedron is obtained by replacing its vertices and faces of the polyhedron. To put it more precisely, the center of each face is regarded as a new vertex, and the centers of the faces sharing the same edge are connected by an edge (therefore the number of edges does not change). The polyhedron obtained in this way is a dual of the original polyhedron. (See also the dual of a graph introduced in Sect. 7.3.) Since each vertex of the original polyhedron can be associated with the center of the newly created face, the dual of the dual returns to the original polyhedron. If you consider dual polyhedra for regular polyhedra, the dual of the regular tetrahedron is itself, the dual of the cube is the regular octahedron, and vice versa, and the dual of the regular dodecahedron is the regular dodecahedron.

Exercise 2.2.2 *Check the dual relationship above by yourself.*

Theorem 2.2.1 *In regular polyhedra, the numbers of edge-unfoldings of duals are the same.*

Proof It is sufficient to show that there is a one-to-one correspondence with each other's unfoldings. We prove it constructively. Consider a cube Q and a regular octahedron Q'. These are mutually dual. Let us consider an edge-unfolding of the cube, let C be the set of cut lines giving the edge-unfolding, and let P be the resulting flat polygon (Fig. 2.2 (1) (2)). (To make it easy to understand, we label each face of the cube, such as T (Top), Bo (Bottom), F (Front), Ba (Back), R (Right), and L (Left). These faces correspond to vertices in the dual regular octahedron. Since the orientation does not change, the relative positional relationship does not change.) On the graph derived from the vertices and edges of Q, C is a spanning tree. Now, consider the connection of the faces of P on the dual graph (Fig. 2.2 (3)). Then, C is a spanning tree on Q and a graph $G(P)$, which represents the connections on the dual graph (thick lines in Fig. 2.2 (3)) also form a spanning tree on the dual Q'. We prove this fact first. Since P is a connected polygon, $G(P)$ is also connected. On the other hand, if $G(P)$ on the dual graph has a cycle, C will be divided into two with the same argument as the proof of Theorem 2.1.1. Therefore $G(P)$ has no cycle. Thus $G(P)$ is always a spanning tree on the dual graph.

Now, the dual graph of Q represents the edges of Q'. On this graph, $G(P)$ is a spanning tree. Thus, $G(P)$ on Q' gives one way to unfold Q' along its edges. For example, the spanning tree of Fig. 2.2 (3) gives the cut lines as in Fig. 2.2(4) on the actual regular octahedron. Thus, by opening up along these cut lines, you can obtain an edge-unfolding in Fig. 2.2(5). Therefore, a one-to-one correspondence is provided in the edge-unfolding of Q and the edge-unfolding of Q'. Thus, the numbers of edge-unfoldings are the same. □

Although the argument itself in the proof of Theorem 2.2.1 is always established between dual polyhedra, it is necessary to be aware that this discussion is *only* considering the method of edge-unfolding. In other words, we do not consider overlap at all here. Therefore, depending on the polyhedron, when you actually open it, the sheet of paper may overlap, and it may not produce a net. In the regular polyhedra, since they have already been confirmed that all the edge-unfoldings do not overlap, one-to-one correspondence between their nets holds true.

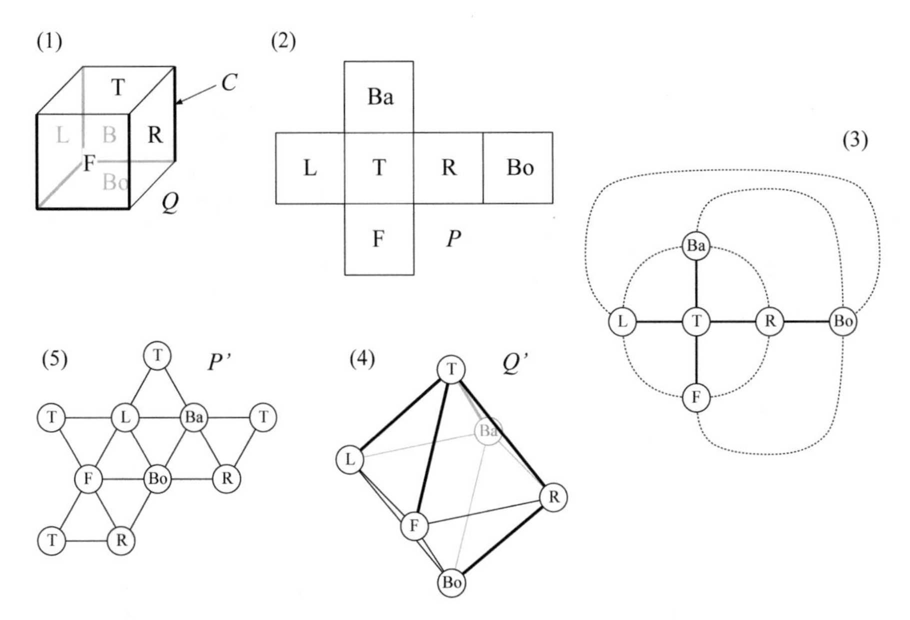

Fig. 2.2 Relationship of edge-unfoldings in dual

2.3 Akiyama-Nara's Theorem

As mentioned in the beginning of this chapter, there are few known facts about general unfoldings. There is only one theorem, which is extremely beautiful. Specifically, it is a very beautiful relationship between unfoldings of a particular family of tetrahedra and certain tiling.

We start with the basics of tiling. A polygon P is *tiling* when you prepare a large number of copies of P, and they make a plane fill without any overlaps and gaps. For example, the following theorem is easy to show.

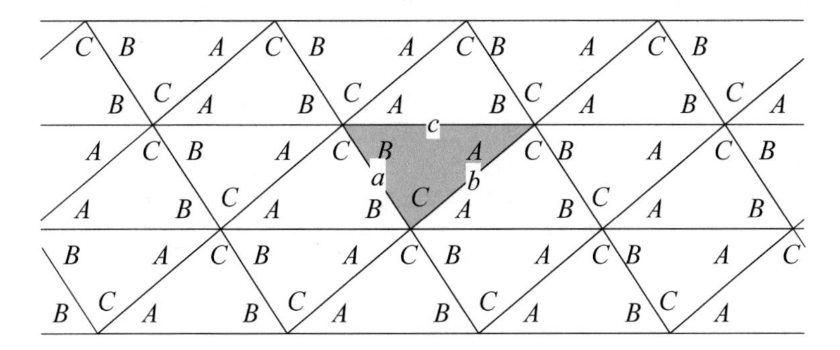

Fig. 2.3 Tiling by arbitrary triangle. For any edge lengths a, b, c, gluing corresponding edges makes the angle $A + B + C = 180°$ so that we obtain a tiling

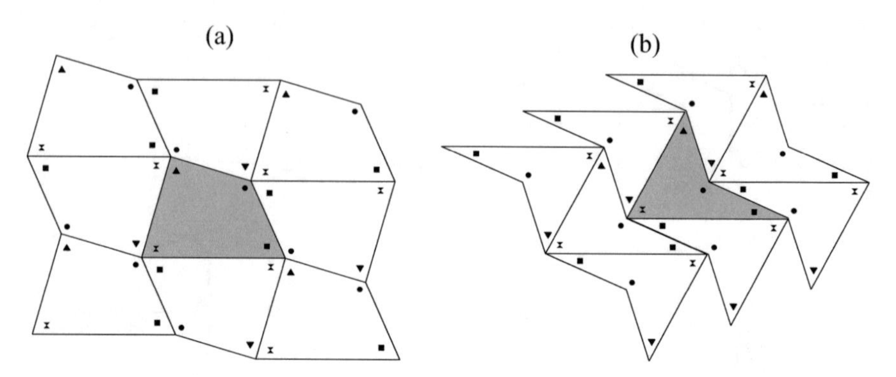

Fig. 2.4 Tiling with any quadrilateral. Convex case (a) and concave case (b). Again, gluing corresponding edges makes the total of four angles 360° so that it becomes a tiling

Theorem 2.3.1 *Any triangle is tiling. Also, arbitrary quadrilateral is tiling.*

Proof Let a, b, c be the lengths of three edges of arbitrary triangle P, and let the angles of opposite corners of each edge be A, B, C. Note that $A + B + C = 180°$ here, and it is easy to fill the plane with copies of P as in Fig. 2.3. Next, consider an arbitrary quadrilateral P'. Note that there are convex and concave quadrilaterals. In both cases, the sum of interior angles is 360° (since both can be divided into two triangles by a (possible) diagonal). If you note this point, when the quadrangle P' is convex, you can fill in the plane with copies as in Fig. 2.4a. When it is concave, it is difficult to imagine a bit, but practically you can fill the plane in the same way as convex (Fig. 2.4b). □

When tiling the arbitrary triangle in the above method, since $A + B + C = 180°$, the obtained tiling is drawn with countless straight lines with no ends. We call the set of lines *general triangular grid* in this book. We usually call the pattern *triangular grid* when the original triangle is a regular triangle. In other words, the triangular grid of a unit regular triangle is generalized to the general triangular grid. The grid obtained by spreading squares in the same way is called *square grid*. (There is a hexagonal grid other than triangular and square grids, but it will not appear in this book.)

Now, if four triangles are suitably selected from the general triangular grid, one larger triangle that is similar to the original triangle can be found. To put it more precisely, it is doubled the length of the edges of the original triangle and is quadrupled the area. If you cut this out and glue the corresponding edges together, it is interesting that a tetrahedron is created (if the original triangle is acute, which is discussed in Sect. 2.3.1). Of course, all faces are congruent. From any acute triangle, you can always make a tetrahedron in this procedure. The tetrahedron which has such four congruent faces is called *tetramonohedron*.

In Theorem 2.3.1, the copies of the polygon P are inverted by 180° and tiled the plane. In particular, we call *p2 tiling* the tiling obtained by the operation of "inverting

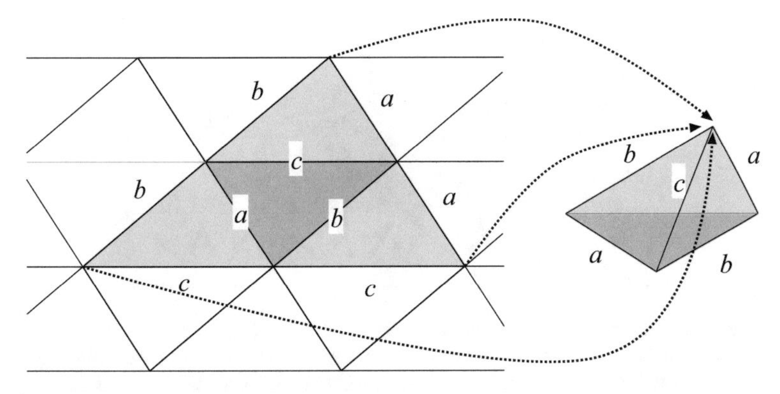

Fig. 2.5 A tetramonohedron and its unfolding

Fig. 2.6 Examples of tilings of the Alhambra. I took hundreds of photos with a few hours' visit

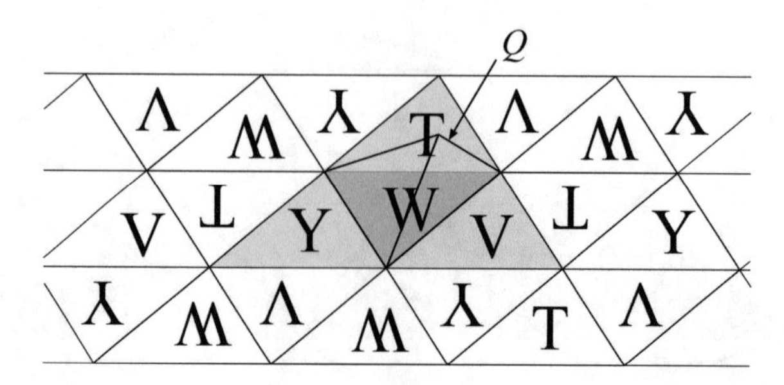

Fig. 2.7 Relationship between tiling and tetramonohedron. Folding the four triangles of gray will result in tetramonohedron Q in the figure. Alphabets are written on each face to distinguish. No matter how you roll the tetramonohedron, when returning to the place where it was visited before, the same face is always on the plane in the same direction

a copy by 180°". (There is a complete classification of tiling with these repeating patterns, and it is known that there are 17 types in total. [p2] is the name of this classification. You can find details of the classification by checking, for example, [GS16]. They say that all 17 types of tiling patterns are used in Spanish Alhambra (Fig. 2.6).) Also, the center point of reversal is called *rotation center*. Assume that the tiling is infinite and the tilings completely overlap when overlapping the two rotation centers, we say that these two rotation centers are *equivalent*.

We show the characterization of a net of a tetramonohedron here.

Theorem 2.3.2 (Akiyama-Nara's Theorem [AN07]) *When the polygon P satisfies the following four conditions, P is a net of a tetramonohedron.*

1. *P is p2 tiling.*
2. *Tiling with P has four kinds of equivalent rotation centers.*
3. *All rotation centers are intersection points on a general triangular grid, and conversely, intersection points on the general triangular grid are all rotation centers of this p2 tiling.*

Looking at P as a net of a tetramonohedron, the points (rotation centers) on the general triangular grid are the four vertices of the tetramonohedron folded from P.

This book gives a proof close to "explanation" which simplified the proof in [Aki07]. The details of the accurate proof can be found in the original paper [Aki07, AN07].

Proof Consider arbitrary tetramonohedron Q. Imagine placing Q on the corresponding p2 tiling (Fig. 2.7). Then roll this Q on the tiling. Then, whichever path you roll, when Q returns to the initial position, Q fits perfectly into the initial position. (This is not the case with cube such as die, and you can have the other face up in any direction. See [BBD+07] for the research of this "rolling cube puzzle".) Here, consider any unfolding P of Q and draw lines to cut along P on Q, that is, the outline of P on Q. We imagine that these lines are a kind of stamp, roll it over the tiling, and *print* the outline of P upon tiling on Q. Then the copies of P shall be infinitely printed on the general triangular grid. Focusing on a triangle of Q, no matter how you roll Q and how many times you visit this triangle, if Q is a tetramonohedron, the same part of the contour of P is printed there. Therefore, P becomes a tiling drawn on the general triangular grid. It would not be difficult to see that the vertices of a general triangular grid correspond to the vertices of tetramonohedron. □

The point is that any tetramonohedra always returns to the original position with the same direction, however, you roll it. Unfortunately, the same argument cannot be established for an unfolding of a cube since the face and direction will change when the cube rolls and returns to its original position in general. Still, there is something we can state. We think about it in Sect. 2.3.3.

By Theorem 2.3.2, it is found that there is a beautiful correspondence between the tiling and nets of a tetramonohedron. Some interesting nets based on the theorem are shown in Fig. 2.8. It will be fun if you do some actual work.

Exercise 2.3.1 *Find the lengths of the edges of each of the tetramonohedra in Fig. 2.8a, c. Also, figure out the lengths of the edges of the box in Fig. 2.8b.*

2.3.1 Tetramonohedron by Acute Triangles

In the previous section, a tetramonohedron plays an important role. We here note that every tetramonohedron has acute triangles as its faces. This fact is a kind of folklore in geometry; however, we state it here in a more precise way.

Theorem 2.3.3 (Folklore) *Let T be a triangle with three edge lengths a, b, c with $a \leq b \leq c$. Then, by the procedure above, four copies of T (1) form a tetramonohedron if T is an acute triangle, (2) form a doubly covered rectangle if T is a right triangle, and (3) do not form any polyhedron otherwise.*

Proof Let $\angle A, \angle B$, and $\angle C$ be the angles at opposite side of the edges of length a, b, and c, respectively. We first prove (3). Without loss of generality, $\angle C > 90°$. Then, since $\angle A + \angle B < 90° < \angle C$, at the apex in Fig. 2.5, these three angles cannot make a closed polyhedron at the apex since the two angles $\angle A$ and $\angle B$ are not enough to cover $\angle C$. Therefore, they cannot make any closed polyhedron without extra folding. (From the other viewpoint, when we try to fold an obtuse triangle into a tetramonohedron in the way in Fig. 2.5, the obtuse angle does not reach any of two acute angles, and these two acute angles do not meet with each other.)

We turn to the case (2) with $\angle C = 90°$. Then the angles $\angle A$ and $\angle B$ exactly fit $\angle C$ since $\angle A + \angle B = \angle C = 90°$. This means that the apex in Fig. 2.5 coincides with the vertex at $\angle C$, the edge of length c is not folded indeed, and we obtain a *doubly covered rectangle*[1] of size $a \times b$.

[1] A *doubly covered rectangle* is a special polyhedron; it consists of two faces of congruent rectangles and four edges, and it has volume zero. In this book, we consider this polyhedron as a degenerated tetramonohedron.

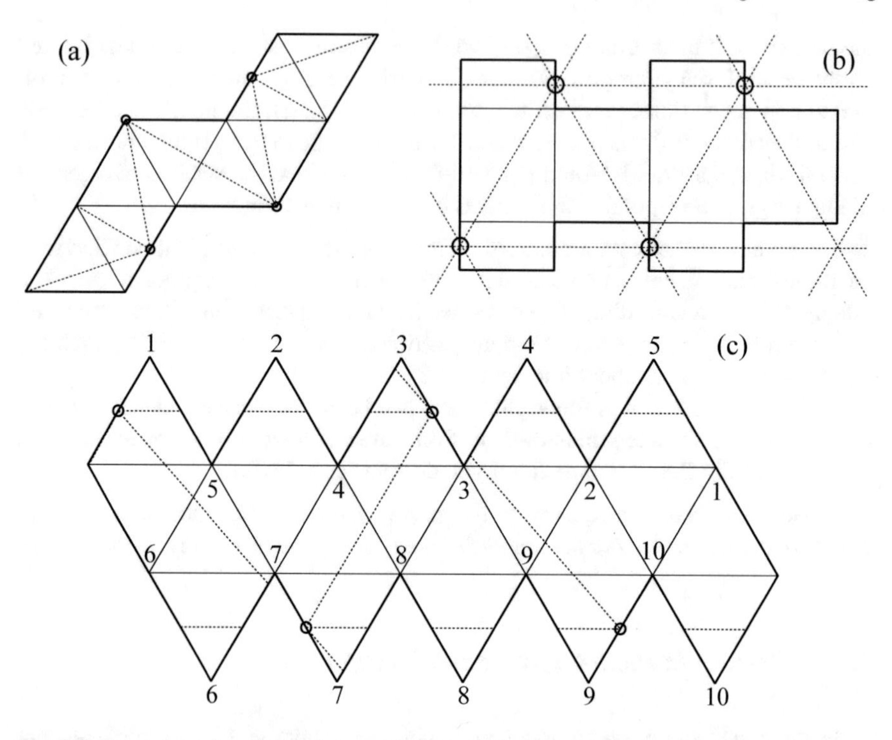

Fig. 2.8 Some examples of nets that are p2 tiling. Each ∘ is a rotation center. (a) Polygon which can be folded to a regular octahedron and a tetramonohedron (Joseph O'Rourke; around 2000?). (b) Polygon which can be folded to a regular tetrahedron and a box (Koichi Hirata, 2004). (c) Polygon which can be folded to a regular icosahedron and a tetramonohedron (Ryuhei Uehara, 2010). When folding along solid lines, a regular octahedron, a box, and a regular icosahedron will be obtained, respectively. On the other hand, folding along the dotted lines, a tetramonohedron, a regular tetrahedron, and a tetramonohedron will be obtained, respectively. In (c), glue the same numbers together

For the case (1), a simple (but indirect) proof can be done as follows. In Sect. 7.2, we will refer to the known formula that gives the volume of any given general tetrahedron characterized by its six edge lengths. In our case, we are given three edge lengths (a, b, c) of the tetramonohedron. Thus it can be folded into tetramonohedron if and only if the formula gives a positive volume. Therefore, we can fold to a tetramonohedron. More rigorous proof will be given in Sect. 7.2.

2.3.2 *Infinite Foldings by Rolling Belt*

In this section, as an example of Theorem 2.3.2, we show unfoldings in which infinitely many different tetramonohedra can be folded. The characteristic named

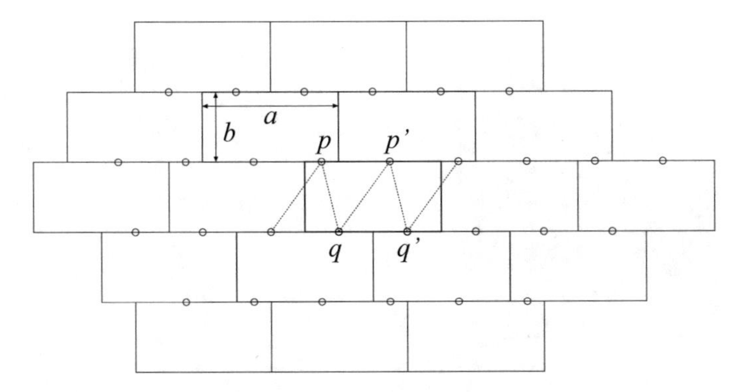

Fig. 2.9 Folding infinitely many tetramonohedra from any rectangle. Each ∘ indicates a rotation center

"rolling belt" appearing in the unfoldings introduced here can be useful when you consider general unfoldings. Nevertheless, the theorem itself is very simple.

Theorem 2.3.4 *From an arbitrary rectangle, infinitely many different tetramonohedra can be folded.*

Proof Let the size of the rectangle be $a \times b$. Then, if you make as in Fig. 2.9, p2 tiling will be completed. Let us state it more precisely. First, place a rectangle. Take the point p arbitrarily on the left half of the upper side. Let p' be the point away from p by $a/2$ to the right. Using this p and p' alternately as the rotation center, rotate the copy of the rectangle by 180° and paste it. Likewise, take the point q arbitrarily on the left half of the lower side and paste the copy as q' the point away from q by $a/2$ to the right. You can fill the entire plane by repeating this alternately with $p, p', q,$ and q'. This is a p2 tiling with these four points p, p', q, q' as the rotation centers and satisfies the conditions of Theorem 2.3.2. Therefore, a tetramonohedron passing through these four points can be folded. Since p and q are arbitrary positions, infinite (so-called uncountable infinite) many types of tetramonohedra can be folded in this way.

In addition, when $b \ll a$, tetramonohedron may not be realized as a polyhedron depending on the positions of p and q; however, the conclusion is unchanged that infinitely many kinds of tetramonohedra are folded. □

There used to be a paper container of drinks such as milk, so-called "tetra-pak". ("Tetra-Pak" itself is the name of an international company, and that container so-called "tetra-pak" is now called "Tetra Classic".) Even today, we sometimes see these tetramonohedron containers (Fig. 2.10). If you know about it, you will take it for granted. If you are not familiar, you can easily realize Theorem 2.3.4 after you actually work as follows (Fig. 2.11).

1. Make a cylinder out of a rectangle.
2. Close the lower part of the cylinder. (Or, if you cut off the bottom part of the envelope, you can get a "bottom closed cylinder" from the beginning.)

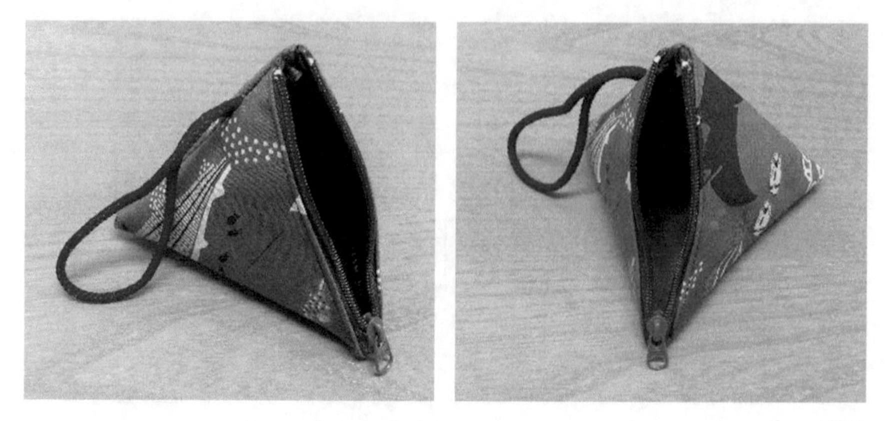

Fig. 2.10 An example of a container using tetramonohedron that can be made with a rectangular sheet. I found this at a souvenir shop in Okinawa, Japan

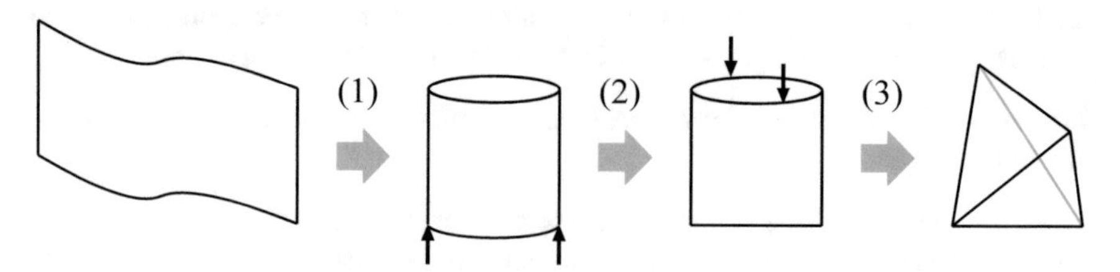

Fig. 2.11 Tetramonohedron made with a cylinder

3. Close the upper part of the cylinder as well. At this time, if the lower closed part and the upper closed part are shifted, a solid can be obtained no matter where it is closed.

In the last step, if you "shift", you always have a tetramonohedron. On the other hand, if you "do not shift", it will be formed of just overlapping two rectangles, namely, a doubly covered rectangle.

Exercise 2.3.2 *Prepare rectangular paper or unnecessary envelope and make tetramonohedra.*

Note that both endpoints of the lower and upper closed parts become the vertices of the resulting tetramonohedron. In other words, six edges of the tetramonohedron are line segments that connect these pairs of two vertices among four vertices, two line segments correspond to the two "closed parts", and four line segments are the other crease lines. The lengths of the two "closed parts" take the same value determined by the size of the rectangle prepared first, which corresponds to the length $a/2$ appeared in the proof of the theorem.

Now think about the closing side, which makes the edge of length $a/2$. Before closing, this side is in the shape of loop, so you can surely compose an edge anywhere you crush and fold. It is also important that the paper gathering around the vertex that is created as a result of folding is exactly 180°. If this is a different angle from 180°, you cannot make such a loop. In such a special circumstance, we can fold in infinitely many ways. From now on, let's call this loop a *rolling belt*. To be more precise definition, the rolling belt is the end of the paper that forms a loop such that any point on the loop makes 180°. When folding a polyhedron from an unfolding, if a rolling belt appears during the folding, you have an infinite way of folding. Namely, you make a vertex (the angle of the surrounding paper at this point is 180°, of course) wherever you like on this loop, make another vertex with the same 180° angle on the opposite side, divide the loop into two halves of the total length, and glue these two line segments into one edge of the polyhedron. This is a fairly unique way of folding and gluing, and it should be noted when discussing the proof about the unfoldings.

2.3.3 Extension of Akiyama-Nara's Theorem to Boxes

There are more opportunities to see boxes than tetrahedra in daily life. We here consider extending Theorem 2.3.2 to the box. It is convenient when considering the unfolding of boxes that appear later in this book.

Theorem 2.3.5 *Consider arbitrary net P of a box Q of size $a \times b \times c$. We assume that a, b, c are natural numbers. (That is, it shall be some numbers multiple of a certain unit length.) Then, if P is placed well on a square grid, all the vertices of Q can be placed on the grid point.*

By Theorem 2.1.3, note that any vertex of Q lies on the boundary of P.

Proof We use the same idea as the proof of Theorem 2.3.2. First, fit the box Q exactly on the square grid of unit size. In other words, make sure that all four vertices of the rectangle face R of the box Q are on the grid point with the correct direction. We roll this Q on the square grid. At this time, the arrangement of the face of Q changes one after another depending on the route to be rolled, but since a, b, c are natural numbers, each vertex of Q always appears on a grid point. Thus, the claim can be proved in the same way as Theorem 2.3.2. Here, consider the net P of Q and draw lines to cut along P on Q, that is, the boundary lines of P on Q. Suppose that the set of these lines is a stamp and we suitably print P on the square grid by rolling Q as an unfolding. Since the vertices of Q are always on the grid points, they are printed on the grid points at the end of the printing of P. □

References

[Aki07] J. Akiyama, Tile-makers and semi-tile-makers. Am. Math. Mon. **114**, 602–609 (2007)
[AN07] J. Akiyama, C. Nara, Developments of polyhedra using oblique coordinates. J. Indones. Math. Soc. **13**(1), 99–114 (2007)

[BBD+07] K. Buchin, M. Buchin, E.D. Demaine, M.L. Demaine, D. El-Khechen, S. Fekete, C. Knauer, A. Schulz, P. Taslakian, On Rolling Cube Puzzles. In CCCG **2007**, 141–144 (2007)
[GS16] B. Grunbaum, G.C. Shephard, *Tilings and Patterns*, 2nd edn. (Dover, 2016)
[HS11] T. Horiyama, W. Shoji, Edge Unfoldings of Platonic Solids Never Overlap. CCCG **2011**, 65–70 (2011)
[HS13] T. Horiyama, W. Shoji, The number of different unfoldings of polyhedra, in *ISAAC 2013*. Lecture Notes in Computer Science, vol. 8283 (Springer, Berlin, 2013), pp. 623–633

Part II
Common Nets

In Part II, we introduce various results on the nets. A net is intuitively a polygon that is cut open with scissors in a polyhedron, but it has to satisfy the following three conditions.

• A net is one connected polygon. That is, it should not be in pieces.
• Each vertex of a polyhedron is definitely opened. That is, a net is flat
• A net must not overlap. In other words, it must be a polygon that can be cut out from a sheet of paper.

On the contrary, it is not necessary to cut along the edges of the polyhedron (as we learned at elementary school). When we only consider the nets made by cutting along the edges of the polyhedron, we say that the nets are obtained by edge-unfolding.

A study on polygons of foldable polyhedra is based on research by Lubiw and O'Rourke in 1996 [LO96]. However, in mathematical terms, it is the current situation that the characterization has not been much found, besides the tiling of Theorem 2.3.2. Conversely, there are lots of problems that cannot be solved unless you actually compute them on a computer, and there is plenty of room for improvement in algorithms. A comprehensive book by Demaine and O'Rourke [DO07, Chap. 25] contains a lot of results related to these polygons and polyhedra foldable from them. Particularly, "one polygon that can fold multiple polyhedra" is an interesting problem that intuition does not work so much. As seen in Fig. 2.8, in general, various polyhedra can be folded from one polygon, but at this time mathematical features are not well known. With a result of several facts, after all, it is often necessary "to check whether all the remaining possible ways can be folded".

Chapter 3
Common Nets of Boxes

Abstract In this chapter, we first consider problems that are easy to handle with computers. Since it seems to be hard to deal with the real coordinate system, polygons on a square grid would be reasonable. Speaking of polyhedra that can be folded from a polygon on a square gird, the first thing that comes to mind is a rectangular parallelepiped, or "box". Is there a single polygon on a square grid that can be folded into multiple rectangular parallelepipeds? The answer is [Yes]. In 1999, Biedl et al. succeeded in actually finding out two polygons that satisfy the following property: "each of two can fold into two different types of boxes" [BCD+99]. (The actual examples are shown in [DO07, Fig. 25.53]. Looking at these two examples, it is hard to make it out that the boxes are folded from it. They said that they found out through trial and error without using a computer.) An example found by me is shown in Fig. 3.1. This example is one of the easiest to understand (by my subjectivity). Are these polygons "exceptional" ones with only a few? Actually, it is not. In this chapter, we introduce the latest results on nets in which these multiple boxes can be folded. Many interesting nets brought by combinations of mathematics and algorithms appear. If you are good at programming, you may want to make it yourself and make new discoveries.

3.1 Some Preparations

Assume that all nets (and each face of the polyhedra) handled here consist of unit squares. (The polygon obtained by gluing the edges of the unit square is called *polyomino*, which is named by Golomb. For fixed sizes are called Monomino (size 1, that is, unit square itself), Domino (size 2), Tromino (size 3), Tetromino (size 4), and Pentomino (size 5). Polyomino of general size n is called n-omino. For details, see [Gol94, Gar08].) In polyomino, we cannot cut or fold on it except at the boundary of the unit squares. Well, here we consider boxes. Without loss of generality, we assume $0 < a \leq b \leq c$ with the lengths of three edges of the box as a, b, c. At this time, the surface area of the box is $2(ab + bc + ca)$. Therefore, for a positive integer S, the set of 3-tuples of integers (a, b, c) satisfying $0 < a \leq b \leq c$ and $ab + bc + ca = S$

R. Uehara, *Introduction to Computational Origami*,
https://doi.org/10.1007/978-981-15-4470-5_3

is denoted by $P(S)$. That is, $P(S) = \{(a, b, c) \mid ab + bc + ca = S\}$. The area of a box can be represented by $2S$. If you try to make a box with a surface area of $2S$, it is obvious that the length of three edges (a, b, c) must be an element in $P(S)$. That is, $|P(S)| \geq k$ is a necessary condition if you fold into k different boxes by one polygon with a surface area of $2S$. (Of course, whether it actually exists or not is the next problem.) For example, the two results of [BCD+99] correspond to $P(11) = \{(1, 1, 5), (1, 2, 3)\}$ and $P(17) = \{(1, 1, 8), (1, 2, 5)\}$. It is easy to make a program that computes $ab + bc + ca$ for each a, b, c that satisfies, say, $1 \leq a \leq b \leq c \leq 50$, and outputs a table for finding the same value. (I indeed made a program that computes $ab + bc + ca$ for a, b, c with $1 \leq a \leq b \leq c \leq 50$, outputs $a, b, c, ab + bc + ca$, sorts them by the value of $ab + bc + ca$, and enumerates all of them. With this computation, recent computers can finish instantly.) According to the output of the program, we can obtain the following results:

$$
\begin{aligned}
P(11) &= \{(1,1,5), (1,2,3)\},\\
P(15) &= \{(1,1,7), (1,3,3)\},\\
P(17) &= \{(1,1,8), (1,2,5)\},\\
P(19) &= \{(1,1,9), (1,3,4)\},\\
P(23) &= \{(1,1,11), (1,2,7), (1,3,5)\},\\
P(27) &= \{(1,1,13), (1,3,6), (3,3,3)\},\\
P(29) &= \{(1,1,14), (1,2,9), (1,4,5)\},\\
P(31) &= \{(1,1,15), (1,3,7), (2,3,5)\},\\
P(32) &= \{(1,2,10), (2,2,7), (2,4,4)\},\\
P(35) &= \{(1,1,17), (1,2,11), (1,3,8), (1,5,5)\},\\
P(44) &= \{(1,2,14), (1,4,8), (2,2,10), (2,4,6)\},\\
P(45) &= \{(1,1,22), (2,5,5), (3,3,6)\},\\
P(47) &= \{(1,1,23), (1,2,15), (1,3,11), (1,5,7), (3,4,5)\},\\
P(56) &= \{(1,2,18), (2,2,13), (2,3,10), (2,4,8), (4,4,5)\},\\
P(59) &= \{(1,1,29), (1,2,19), (1,3,14), (1,4,11), (1,5,9), (2,5,7)\},\\
P(68) &= \{(1,2,22), (2,2,16), (2,4,10), (2,6,7), (3,4,8)\},\\
P(75) &= \{(1,1,37), (1,3,18), (3,3,11), (3,4,9), (5,5,5)\}
\end{aligned}
$$

For example, a net that folds into two boxes has an area of 22 or more if possible, and a net that folds into three boxes, if present, has an area of 46 or more, and so on. When the sizes of two boxes are (a, b, c) with $a \leq b \leq c$ and (a', b', c') with $a' \leq b' \leq c'$, we define these boxes are *different* if $a \neq a'$ or $b \neq b'$ or $c \neq c'$ holds.

In general, as we increase the value gradually, it seems that surface area with many possible combinations appears. The above data also seems to support it. This intuition is correct. This is somewhat skillful, but we show the following theorem.[1]

Theorem 3.1.1 *For an arbitrary natural number p, there is an area common to at least p different boxes.*

[1]This theorem and proof are pretty clever ideas by Mr. Toshifumi Okumura who belonged to my laboratory in 2014.

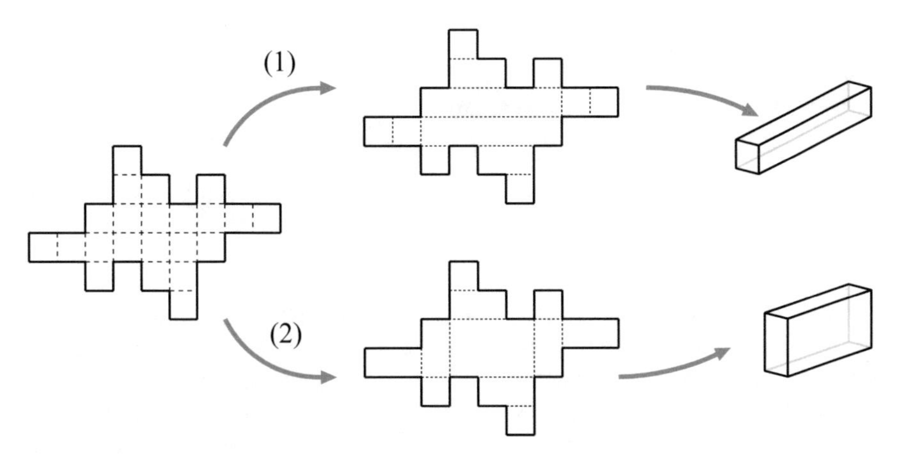

Fig. 3.1 A net of two boxes. By folding along the crease lines of (1), the box of size $1 \times 1 \times 5$ can be obtained, and by folding along the crease lines of (2), the box of size $1 \times 2 \times 3$ can be obtained

Proof For the given p and each natural number $i = 1, 2, \ldots, p$, let $a_i = 2^i - 1, b_i = 2^{2p-i} - 1$, and $c_i = 1$. Then, for any i, $a_i b_i + b_i c_i + c_i a_i = (2^{2p} - 2^i - 2^{2p-i} + 1) + (2^{2p-i} - 1) + (2^i - 1) = 2^{2p} - 1$. Moreover, (a_i, b_i, c_i) and (a_j, b_j, c_j) are obviously different for any $1 \leq i < j \leq p$, which completes the proof. □

From Theorem 3.1.1, if we consider only the area, we cannot deny the possibility that a common net exists for any number of different boxes.

Let B be a box whose three edge lengths are a, b, c. Then B has two rectangular faces of size $a \times b, b \times c$, and $c \times a$. Each rectangle is considered a collection of unit squares. That is, B is composed of $2(ab + bc + ca)$ unit squares. Then we define the *dual graph* $G(B) = (V, E)$ of B as follows. The elements of vertex set V are the unit squares. That is, V has $2(ab + bc + ca)$ elements. Two unit squares u and v in V are joined by an edge $\{u, v\} \in E$ if and only if u and v share an edge on B, that is, they are adjacent on B. Thus $G(B)$ is a four-regular graph consisting of $2(ab + bc + ca)$ vertices since each unit square is always adjacent to four different unit squares. Therefore, $|E| = 4(ab + bc + ca)$ is established. The following lemma is obtained for this graph $G(B)$.

Lemma 3.1.2 *Let T be the spanning tree of $G(B)$ for a given box B. For each edge $\{u, v\}$ not in T, cut along the edge shared by the unit squares u and v on B. Then we can obtain an unfolding P corresponding to T in B.*

Proof Since T is connected, P never be disconnected after cutting edges. Also, the corner part of B produces a closed cycle of length 3 on $G(B)$, but since T does not include any cycle, the corner should be opened on P. Thus P is an unfolding of B.□

The polyomino P appearing in Lemma 3.1.2 is an "unfolding". Is it indeed a "net", which is a simple flat polygon that can fold into the box? In general, it is "Yes" but there are two points to keep in mind.

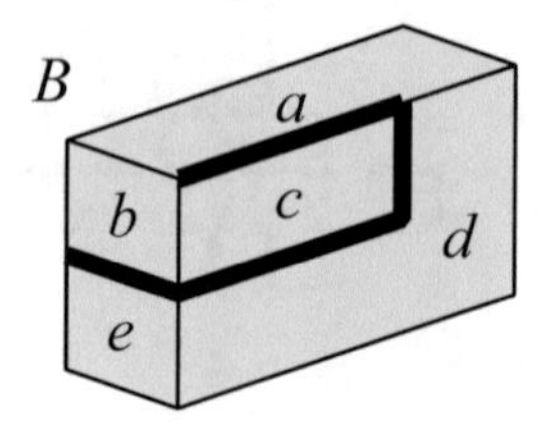

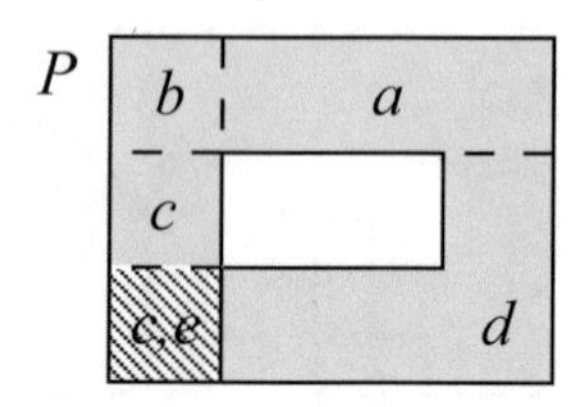

Fig. 3.2 An example of cutting open and overlapping of a (part of) box

First, if P has an overlap, it is not a net by the definition. However, does a box originally cut open with straight lines parallel to the edges and overlap each other? Surprisingly, such an example can be found relatively easily. Considering the unfolding P shown in Fig. 3.2, although this is a part of an unfolding of a box, it has an overlap. Therefore, such a cut unfolds a box, but the result is not a net. In addition, if you cut off between d and e instead of b and e in the figure, you can obtain an unfolding that "touches".

In this way, we call *touch* if two unit squares that were not adjacent to each other on the original box are adjacent to each other when they are unfolded. If these unit squares coming around from different directions come into touch with each other, an apparent "hole" may be emptied during unfolding. There is no definite rule whether to call such polygons with "holes" as nets. Since it is not overlapped, it might be natural to call them nets. However, when actually solving the problem, it is hard to deal with. Therefore, in this book, we will not handle polygons with such apparent holes as nets. In the term of computational geometry, we only consider simple polygons without (apparent) holes. I present one problem related to this.

Open Problem 3.1.1 *The reason why this book does not handle with a polygon with a hole as a net is to simplify the argument (and, in fact, it is fruitfully enough). When folding a box, for example, with a polygon with a hole, you must cut somewhere to break the hole. Actually, if you fold a polygon with a hole as it is, you cannot fold any convex polyhedron. This may not be obvious, but it can be derived from Lemma 3.1.3. Therefore, if there is a hole, it must be surely cut somewhere and a box will not be folded from it without connecting to the outer circumference. Speaking of the topic "polygons foldable into multiple boxes" dealt with this chapter, it is not known whether or not different boxes can be folded in two or more different cutting ways for given polygons with holes.*

Hereafter, it is assumed that P, which is a polygon obtained by unfolding a box B, has neither an overlap nor (apparent) a hole. That is, the net P is identified by the spanning tree T that gives a cut of B. However, P sometimes contains redundant cuts. An example is shown in Fig. 3.3. In this case, the cut line in the original box B is redundant as it is not useful for unfolding even on the original B or the net P. That is, this cut is not required. Such redundant cuts can be characterized by the following lemma.

Lemma 3.1.3 *Let P be a polygon that can fold into a box B. Suppose P contains two unit squares with cut between them like A and D in Fig. 3.4. More precisely, we sup*

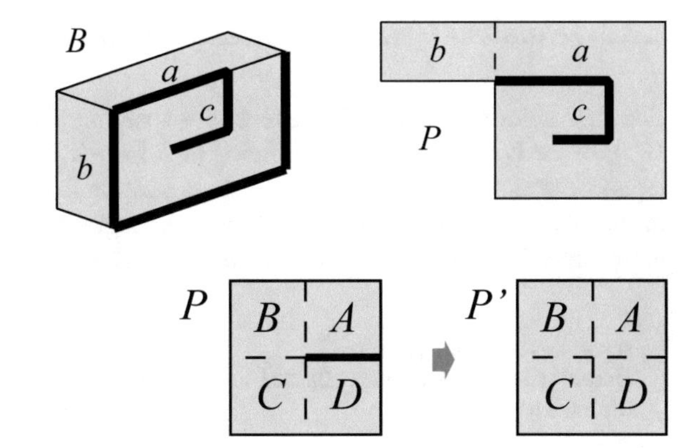

Fig. 3.3 Unnecessary cut

Fig. 3.4 Glueing

pose that the edges on B corresponding to the three edges of the cycle (A, B, C, D, A) *of length 4 included in* $G(B)$ *(the edges between A and B, between B and C, and between C and D in the figure) are not cut on* P*, and only the edge on B corresponding to one edge (between A and D in the figure) has been cut. Let* P' *be a polygon that is obtained from P by gluing this pair (A and D in the figure). Then,* P' *can also fold into* B*.*

Proof Since B is a closed convex polyhedron, assuming that P' does not fold into B, another unit square must be placed between A and D. At this time, however, there exists a vertex where more than 5 unit squares gather around. With this vertex, the orthogonal convex polyhedron B cannot be folded. Therefore, there is no extra unit square between A and D on B, and you can make the same polyhedron as P by gluing A and D. □

You can apply "glueing" of Lemma 3.1.3 repeatedly. Specifically, you can eliminate redundant cuts in P by repeating and removing "cuts that end in the face on B".

Note that the proof of Lemma 3.1.3 uses the convexity of a box. If the polyhedron is not convex, we cannot determine if a cut is redundant or not in general. In fact, recently, it is proved that a rectangle with some slits can fold to any number of different polycubes if it is sufficiently large [DHK+19].

3.2 Common Nets of Two Boxes

Several algorithms can be considered to generate common nets in which two boxes can be folded. My research group has implemented three algorithms so far. In this book, we outline each algorithm and introduce a number of interesting nets obtained as results.

3.2.1 Random Generation Method of Spanning Trees

The first algorithm randomly generates a spanning tree for the given dual graph $G(B)$ of the box B, creates an unfolding of B from it, and converts it to an integer to register/inspect using a huge hash table. Repeating this process for the set of boxes of the same surface area for a long time, the algorithm will find a common net if these boxes have. The outline of the algorithm is as shown in Fig. 3.1.

Input : S satisfying $|P(S)| > 1$;
Output: Common nets of area $2S$ that can fold into two boxes;
Clear the hash table H;
while *true* **do**
  Randomly choose an element $t = (a, b, c)$ from $P(S)$;
  Randomly generate a spanning tree of the dual graph $G(B)$ for the box B of size $a \times b \times c$;
  Represent unfolding P corresponding to the spanning tree T in a 0/1 matrix;
  if (t', P) *with* $t \neq t'$ *is in* H **then** Output P (with corresponding edge lengths);
  if P *is not yet registered in* H **then** Register (t, P) into H;
end

Algorithm 3.1: Algorithm for generating common nets using spanning trees.

In this algorithm, polygon P is represented with 0/1 matrix in a natural form. That is, we put P on a sufficiently large array, and the place where there is a unit square is set to 1, and the place where there is no unit square is set to 0. (We indeed trim the array to make P inscribe to the rectangular array.) In this way, there are the following two nice properties.

- Redundant cut lines are not automatically reduced.
- The number of 1 is $2S$, the unfolding does not overlap, or it is a net.

That is, if the number of 1 is properly $2S$, and there is no hole, it turns out that this is a correct net.

When registering the net in the hash table, you also need to be careful about the orientation of the net. There are eight ways to layout the net when we consider rotation and flipping. In the algorithm, we generate all eight types and bring them to "upper left" as much as possible. Then we use the lexicographically smallest one as the "standard form" of this net. That is, every time a single net is generated, it is replaced with the standard form and registered in the hash table on it.

When I designed this algorithm, since the purpose of this algorithm was to find a net shared by two or more boxes, I cut corners with the implementation. Specifically, although there are the following two drawbacks, I did not cope with it on purpose. First, random generation of the spanning tree is not necessarily uniformly generated because it "repeatedly chooses an edge randomly and adds it as far as it is a tree". Considering unnecessary cutting, intuitively, it is more likely to be generated as a net having a larger face. Also, I did not check the presence of holes because I thought that overlapping and holes would not be generated so much. While executing, when there was an overlap, the program reported a message to that issue but overlaps occurred in about five percent of the trials. In addition, although 2165 outputs were obtained by this program, 26 were not recognized as solutions because of the hole, and the remaining 2139 satisfied the condition of nets.

From an algorithmic point of view, there are algorithms that check if a given polygon has a hole in linear time. For example, an algorithm by Asano and Tanaka can solve this problem in linear time using only constant memory [AT08].

3.2.1.1 Experimental Results of the First Algorithm

When I ran the algorithm that generated a spanning tree with a normal laptop computer (IBM ThinkPad X40: 1 GB CPU with 1.5 GB of memory), it could generate about 3×10^6 nets in 1 h and output about 100 solutions to $P(11)$. Thus, I used a supercomputer (SGI Altix 4700: 96 CPU with 2305 GB of memory). For generating random numbers, I used the Mersenne-Twister method.[2] Experimental results are shown in Table 3.1. In the table, "$2S(S)$" is the area of the polygon and its half, "$|P(S)|$" is the number of possible boxes, "Number of generation" is the number of trials, "Box size" is the size of the boxes, "Number of solutions" is the number of obtained nets, and "Error" is the number of developments with holes. For example, in the column of $P(11)$, the algorithm generated about 6.7×10^7 nets for the box of size $(1, 1, 5)$ or $(1, 2, 3)$, and 556 polygons were obtained. Among them, 15 are polygons with holes, and the remaining 541 polygons are common nets of the box of size $(1, 1, 5)$ and the other box of size $(1, 2, 3)$. In total, I obtained 2139 simple nets that can fold into two boxes. Programs for various parameters were executed appropriately in parallel on a supercomputer and were executed for several days to several weeks. Then, when each process is using the memory too much, execution was stopped as appropriate. A part of the obtained solutions is shown in Fig. 3.5. All the solutions obtained by this algorithm are published at http://www.jaist.ac.jp/~uehara/etc/origami/nets/index.html.

3.2.2 *Direct Search Algorithm for Common Nets*

The second algorithm first determines the set of plural boxes having the same surface area as the target. Let B_1 and B_2 be two boxes in this set. Then the algorithm determines uniformly at random a unit square s_1^1 on B_1 and a unit square s_2^2 on B_2. These unit squares have their "directions" determined at random. The two squares s_1^1 and s_2^1 are assumed to correspond to the same square of the same direction on the common net. The algorithm will "grow" the common partial net P of these B_1 and B_2 simultaneously by the following procedure: First, it checks the outer perimeter e of the current P. Let s be the unit square located next to P on edge e (not overlapping

[2] http://www.math.sci.hiroshima-u.ac.jp/~m-mat/MT/emt.html.

Table 3.1 Experimental results of the first algorithm

$2S(S)$	$\|P(S)\|$	Number of generations ($\times 10^7$)	Box size	Number of solutions	Error
22(11)	2	6.7	(1, 1, 5), (1, 2, 3)	541	15
30(15)	2	18.6	(1, 1, 7), (1, 3, 3)	72	1
34(17)	2	28.4	(1, 1, 8), (1, 2, 5)	708	0
38(19)	2	30.4	(1, 1, 9), (1, 3, 4)	41	0
46(23)	3	191.0	(1, 1, 11), (1, 3, 5)	568	3
			(1, 2, 7), (1, 3, 5)	92	5
54(27)	3	126.7	(1, 1, 13), (3, 3, 3)	2	0
			(1, 3, 6), (3, 3, 3)	1	0
58(29)	3	89.3	(1, 1, 14), (1, 4, 5)	37	0
62(31)	3	82.4	(1, 3, 7), (2, 3, 5)	5	0
64(32)	3	204.8	(1, 2, 10), (2, 2, 7)	50	2
			(2, 2, 7), (2, 4, 4)	6	0
70(35)	4	91.3	(1, 1, 17), (1, 5, 5)	3	0
			(1, 2, 11), (1, 3, 8)	11	0
88(44)	4	217.0	(2, 2, 10), (1, 4, 8)	2	0
90(45)	3	34.6		0	0
94(47)	5	51.3		0	0
112(56)	5	36.0		0	0
118(59)	6	35.5		0	0
Total	–	–	–	2139	26

P). Then the algorithm can identify the corresponding unit square s' on B_1 and the corresponding unit square s'' on B_2 to s. If they are not yet belonging to P on both of B_1 and B_2, it is said that the edge e is *extensible*. The algorithm examines the extensible edges of P, picks e from them uniformly at random, and adds the unit square s to P at the position of e. This process stops in the following two cases. If P contains all the unit squares and becomes the same area as the surface area of B_1 (and B_2), P is the desired net so the algorithm can output it and halt. On the other hand, if there are no more extensible edges before the first case, the algorithm fails to find the solution. The detail of the algorithm is shown in Algorithm 2.

This algorithm can be easily extended to search for three or more boxes. Also, when an edge is extensible, it is guaranteed not to overlap, so there is no need to check an overlap.

3.2.2.1 Experimental Results of the Second Algorithm

The second algorithm was run on a laptop computer (Intel Core 2 Duo CPU T9900 3.06 GHz Windows 7). For each size, the program was executed for 1×10^7 times.

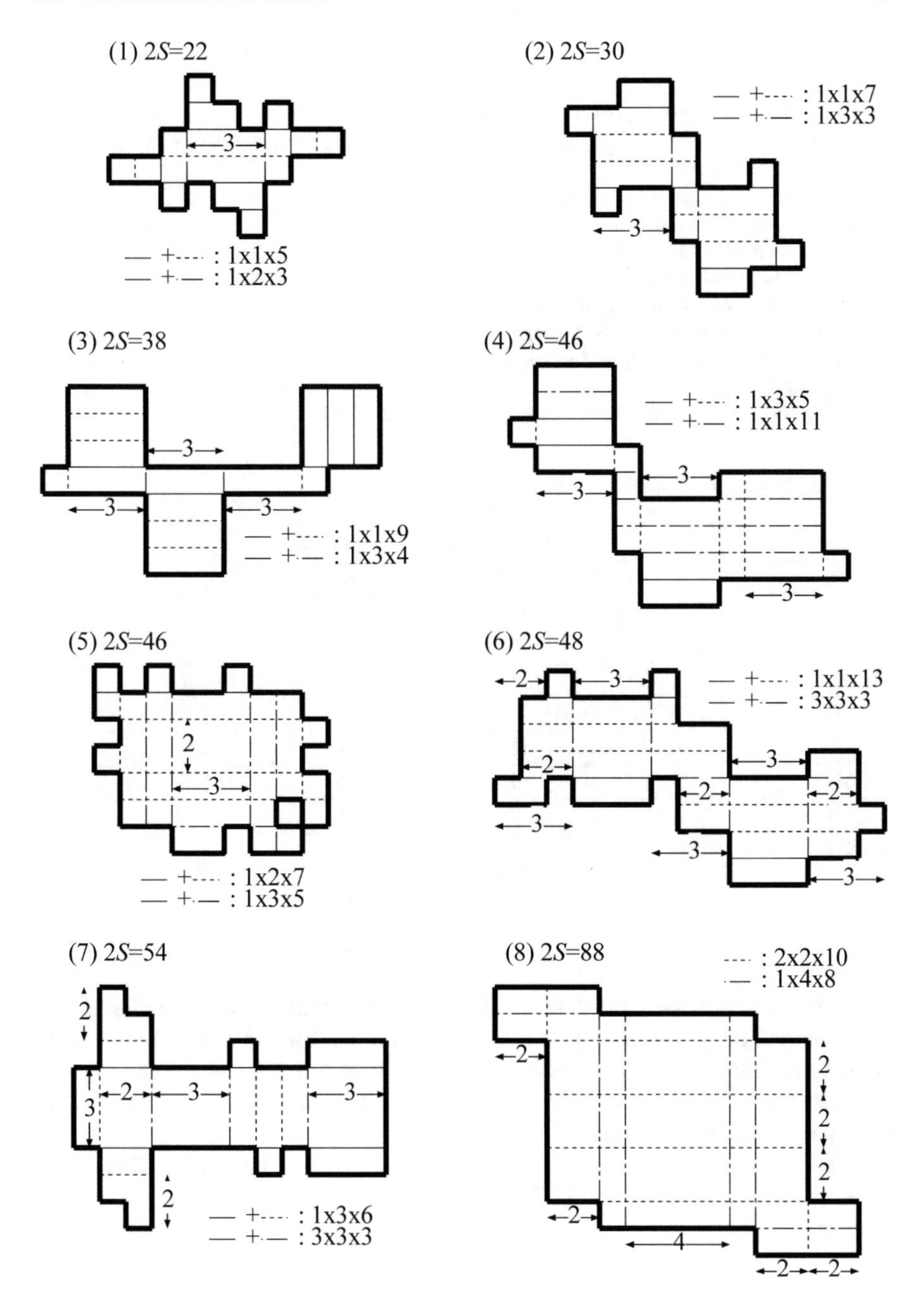

Fig. 3.5 A part of the obtained solutions

Input : S satisfying $|P(S)| > 1$;
Output: Polygon of area $2S$ that can fold into two boxes;
Choose two types $t_1 = (a_1, b_1, c_1)$ and $t_2 = (a_2, b_2, c_2)$ from $P(S)$;
Choose a unit square s_1^1 on the box B_1 of type t_1 and a unit square s_2^1 on the box B_2 of type t_2 uniformly at random;
Initialize P by a unit square corresponding to s_1^1 and s_2^1;
repeat
  For each edge e of P, check if e is extensible or not;
  if *no extensible edge* **then** output "fail to find" and halt;
  Choose an extensible edge e uniformly at random;
  Add a unit square s to P at the edge e;
until *all unit squares in B_1 and B_2 are joined to P*;
Output P.

Algorithm 3.2: Algorithm that generates a common net simultaneously.

Table 3.2 Experimental results of the second algorithm

$2S(S)$	Box size	Computation time (s)	Number of solutions	Error
22(11)	(1, 1, 5), (1, 2, 3)	100	1863	13
30(15)	(1, 1, 7), (1, 3, 3)	155	370	0
34(17)	(1, 1, 8), (1, 2, 5)	191	3705	1
38(19)	(1, 1, 9), (1, 3, 4)	213	914	0
54(27)	(1, 1, 13), (3, 3, 3)	352	690	1
54(27)	(1, 1, 13), (1, 3, 6)	351	717	1
54(27)	(1, 3, 6), (3, 3, 3)	477	243	3
88(44)	(1, 4, 8), (2, 2, 10)	1007	153	0
88(44)	(2, 2, 10), (2, 4, 6)	967	86	1
Total	–	3813	8741	20

The experimental results are shown in Table 3.2. Each entry conforms to Table 3.1. For example, the program tried 1×10^7 times for $P(17)$, and found 3705 simple common nets and 1 non-simple common net for a box of size $1 \times 1 \times 8$ and a box of size $1 \times 2 \times 5$. The execution time was 191 s. We were able to find a total of 8741 common nets that fold into two different boxes.

In particular, a graph of the number of trials for $P(11)$ and the number of solutions found in the trial is shown in Fig. 3.6. Observing this graph, we can expect that the number of common nets of two boxes of size $1 \times 1 \times 5$ and $1 \times 2 \times 3$ will be roughly 2000 or so. As a matter of fact, this number was 2263 in total.

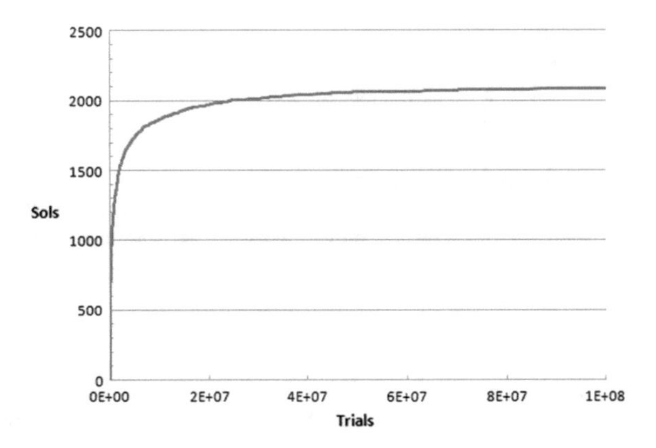

Fig. 3.6 Relationship between the number of trials and the number of solutions for $P(11)$

3.2.3 Brute Force Search Algorithm

Since the algorithms in Sects. 3.2.1 and 3.2.2 unfold boxes randomly, even if the boxes have a certain large area, you may find some common nets if you carry them on for a long time. On the other hand, if the area is limited to small, you may search entirely. Especially, in the smallest area 22 where there is a common net of two boxes, we have found all common nets of two boxes of size $1 \times 1 \times 5$ and $1 \times 2 \times 3$. Specifically, there are 2263 common nets. (Since it is too much to publish in this book, they are listed at http://www.jaist.ac.jp/~uehara/etc/origami/nets/all-22.html.)

The basic idea of this algorithm is *breadth-first search (BFS)* in the algorithm theory. Now consider a simple polygon P' obtained by removing a part of P which is a net of a box. That is, P' is completely contained in P, and it is a simple polygon formed by unit squares. Then, of course, P' can "paste" onto this box without overlap. We call such a polygon *partial net*. For example, one unit square can be considered as a common partial net of all boxes. The same applies to a rectangle of size 1×2 connecting two unit squares. In this section, we consider these common partial nets limited to boxes of size $1 \times 1 \times 5$ and of size $1 \times 2 \times 3$. Let denote by L_i a set of common partial nets of area i for these two boxes. Then $|L_1| = |L_2| = 1$ and $|L_3| = 2$. We will expand these sets in order. In other words, our algorithm computes L_i from L_{i-1}. Specifically, Algorithm 3 was implemented and executed.

This scale of algorithm can execute with a standard PC. The running time was approximately 10 h for implementation in 2011 and about 5 h in 2014. A table summarizing the experimental results is shown in Table 3.3. When i is small, most i-ominoes are common partial nets, but most of them are dropped in the end. Also, we can observe that the common partial nets of area 18 are the largest number, and the computation per area from 16 to 19 is a bottleneck.

Input : None;
Output: The whole common nets of two boxes of size $1 \times 1 \times 5$ and $1 \times 2 \times 3$;
Let L_1 be a set of one unit square;
for $i = 2, 3, 4, \ldots, 22$ **do**
  $L_i := \emptyset$;
  for *each common partial net* P *in* L_{i-1} **do**
    for *each polygon* P^+ *obtained by attaching one unit square to* P **do**
      Register P^+ into L_i when P^+ is a common partial net of two boxes and not yet registered in L_i;
    end
  end
end
Output L_{22};

Algorithm 3.3: BFS algorithm.

Table 3.3 The numbers of common partial nets of two boxes of size $1 \times 1 \times 5$ and $1 \times 2 \times 3$. For comparison, the numbers of i-ominoes are also noted

i	1	2	3	4	5	6	7	8	9
$\|L_i\|$	1	1	2	5	12	35	108	368	1283
i-ominoes	1	1	2	5	12	35	108	369	1285

i	10	11	12	13	14
$\|L_i\|$	4600	16388	57439	193383	604269
i-ominoes	4655	17073	63600	238591	901971

i	15	16	17	18
$\|L_i\|$	1632811	3469043	5182945	4917908
i-ominoes	3426576	13079255	50107909	192622052

i	19	20	21	22
$\|L_i\|$	2776413	882062	133037	2263
i-ominoes	742624232	2870671950	11123060678	43191857688

3.3 Interesting Common Nets

In Sect. 3.2, we considered the nets common to two boxes, but a number of interesting nets were obtained in the research process. We introduce them in this section.

3.3.1 *Tiling Net*

As we have seen in Sect. 2.3, there exists a relationship between net and tiling. There is also a merit that tiling does not waste when cutting out. Then does tiling exist in such common nets? The tiling found by me from the output of the first algorithm is shown in Fig. 3.7. This is a common net of a box of size $1 \times 1 \times 8$ and another

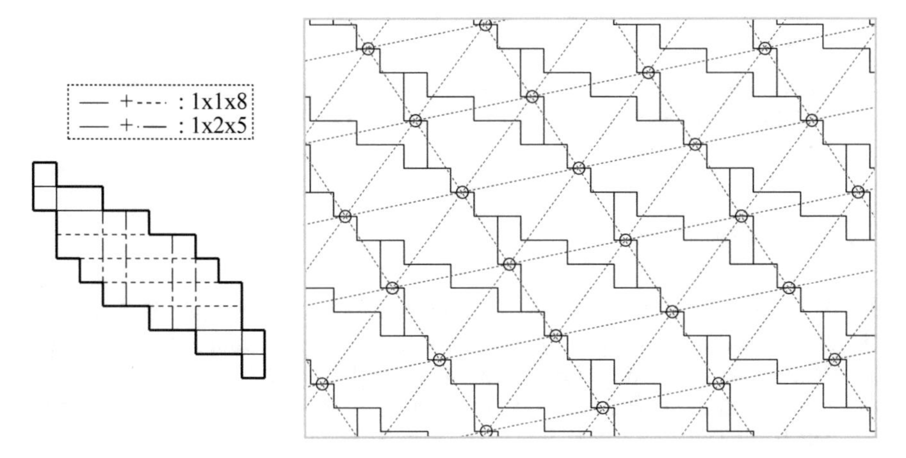

Fig. 3.7 A p2-tiling that can fold into two boxes of size $1 \times 1 \times 8$ and $1 \times 2 \times 5$. If you fold as shown on the left in the figure, you can fold into one of two boxes. On the other hand, if you fold as shown on the right in the figure, you can fold into a tetramonohedron. The circles in the right figure are rotation centers of the p2-tiling

box of size $1 \times 2 \times 5$. Moreover, it is also p2-tiling; hence, we can fold it into a tetramonohedron as shown in the figure!

Kano et al. named a figure (to be exact, a pair of a solid and its net) with the properties that the net fills the plane and the solid folded from it embeds the space *double packable solid* in [KRU07, Sect. 3. 5. 2]. From this point of view, since arbitrary box fills space, Fig. 3.7 gives two types of double packable solids at the same time.

3.3.2 Net Where Crease Lines Do Not Cross

When folding a polygon to make a box from two possible candidates, engineeringly, it may be desirable that crease lines do not cross each other. Is there a net with such a characteristic? The answer is [Yes]; Fig. 3.5(3), (7) satisfy this property.

3.3.3 Net Where Crease Lines Are Independent

When folding a polygon to make a box from two possible candidates, it may be desirable that crease lines are independent of each other in each folding way, or it is preferable that they share no common crease lines. Does a net with such a characteristic exist? The answer is also [Yes], and Fig. 3.5(8) is a net that satisfies this property.

We have not yet explored further properties:

Open Problem 3.3.1 *Is there any common net that can fold into two boxes such that their crease lines are independent and they do not cross each other?*

3.3.4 Infinite Nets

In this section, we show that there are infinitely many different nets that can fold into two boxes. (Here, for given two boxes of size $a \times b \times c$ and $a' \times b' \times c'$, they are *different* boxes when $gcd(a, b, c, a', b', c') = 1$ is established. To put it a little more, we avoid the trivial nets such as a net of a box of size $(2a) \times (2b) \times (2c)$ obtained from a net of a box of size $a \times b \times c$.) Utilizing the nets already provided, the following generalization theorem is obtained.

Theorem 3.3.1 *For any two positive integers j and k, there exists a common net of two boxes of size $1 \times 1 \times (2(j+1)(k+1)+3)$ and $1 \times j \times (4k+5)$.*

Proof For any given positive integers j and k, we show that the polygon in Fig. 3.8 satisfies the condition. The way of folding each box is the same as the way of folding one in Fig. 3.7. Since the initial parameter j determines only the width of the rectangle in Fig. 3.8, there are no effects on the way of the folding of these two boxes. For the second parameter k, copy the polygonal area on the left side of Fig. 3.8, paste it so that the gray area overlaps, and repeat this "copy and paste" for k times. Then, there are two kinds of folding ways for each k. The first way is essentially the same for all k; fold four unit squares so that they are rolled in the vertical direction, and we can obtain a box of size $1 \times 1 \times (2(j+1)(k+1)+3)$. On the other hand, in the second way, we fold the whole body horizontally according to k to twist it in a spiral of k times. This way of folding gives us a box of size $1 \times j \times (4k+5)$. □

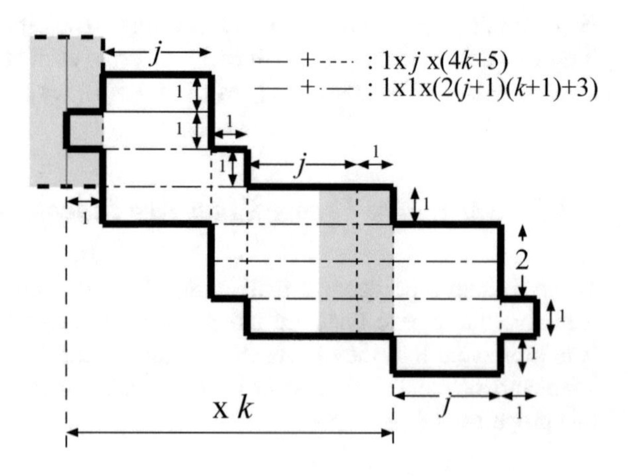

Fig. 3.8 A common net of two boxes of size $1 \times 1 \times (2(j+1)(k+1)+3)$ and $1 \times j \times (4k+5)$

We here note that in Theorem 3.3.1, k should be a positive integer; however, j can be any positive real number. Therefore, Theorem 3.3.1 implies the following corollary.

Corollary 3.3.2 *There are uncountable infinitely many nets that can fold into two different boxes.*

Note that, in Sect. 3.4.2, we also show that there are infinitely many nets that can fold into three different boxes in another way (by Theorem 3.4.2).

3.3.5 More General Problems

In this section, we show the further results based on the obtained research on "common nets of two boxes". In the following discussion, although polyhedra other than boxes appear, all of them are formed by unit squares. Moreover, the angle of each fold is a multiple of 90°. We call these polyhedra *orthogonal polyhedra*.

First, since we have $|P(S)| \leq 1$ when $S < 11$, there are no orthogonal nets that can fold into two different boxes of area less than 22. However, if you allow the "same" boxes, there is a smaller solution. Figure 3.9 is that, and you can fold into a box of the same type in three different ways (bold lines are gluing lines). It is well known that there are eleven different nets in a unit cube by edge-unfolding (Fig. 1.1), and each of them is folded in a unique manner. Therefore, Fig. 3.9 is the net with the smallest area where multiple folding ways exist.

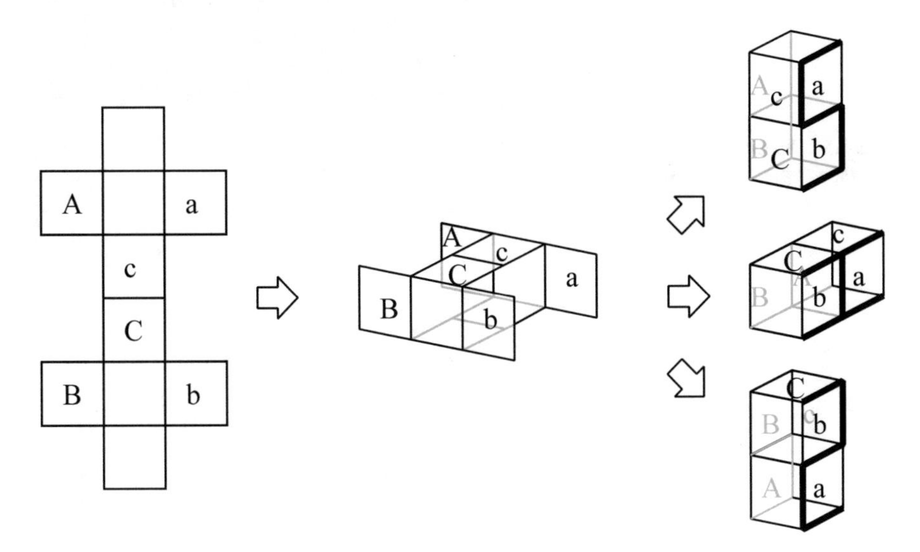

Fig. 3.9 A net with three different ways of folding

Next, we consider common nets of plural non-isomorphic orthogonal polyhedra which are not necessarily convex. In this case, it is already known that there are some polyominoes of area 18 which are common nets of seven different orthogonal polyhedra (which are also all possible ones of area 18). One of them was devised by George Miller and Donald E. Knuth and is sold as a puzzle named "Cubigami 7" (Figs. 3.10 and 3.11). This puzzle won an honorable mention at the 2005 World Puzzle Contest (for details, see [MK06]).

In Lemma 3.1.3, we showed that there is no slit that ends in the face, but it was used in the proof that the objective polyhedron is convex. In Cubigami 7, some objective polyhedra are not convex; therefore, there are some slits that end in the face of this common net. According to the analysis of Knuth, there are 68 simple common nets that fold into these seven orthogonal polyhedra. By the program which remodeled

Fig. 3.10 Cubigami 7

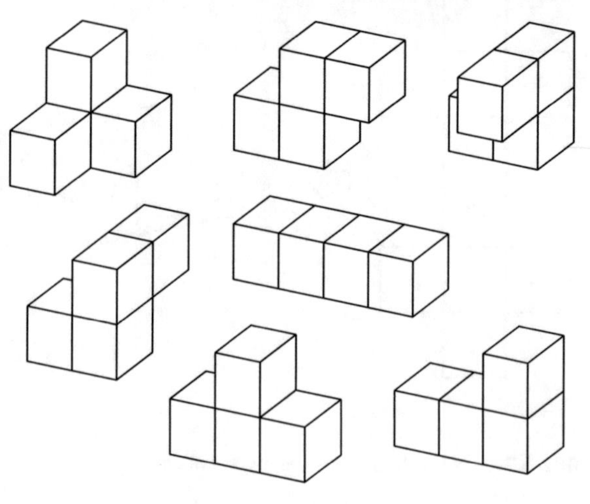

Fig. 3.11 Seven different orthogonal polyhedra of area 18

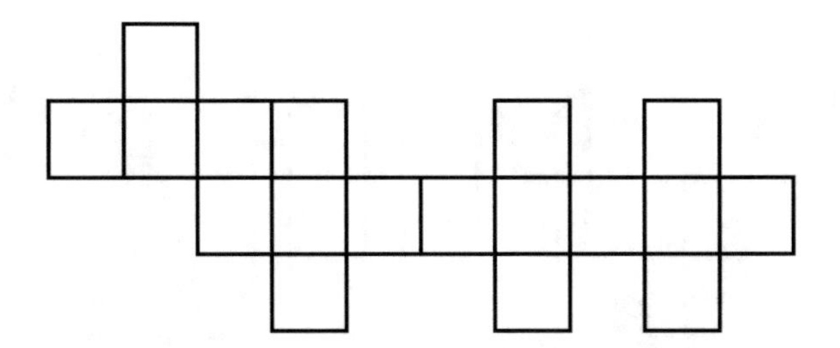

Fig. 3.12 A simple common net that folds seven different orthogonal polyhedra

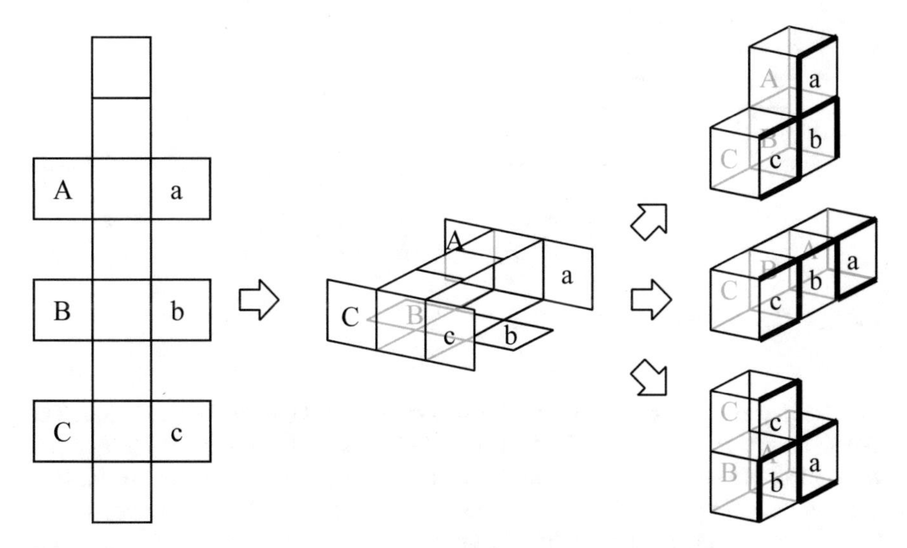

Fig. 3.13 A net in which orthogonal polyhedra can be folded in three different ways

the second algorithm of this chapter, we also found one solution by trials of 1×10^8 times (Fig. 3.12).

In the meaning of "a common net that can fold into all orthogonal polyhedra for a given area", Cubigami 7 is not a solution to the smallest area. A net in Fig. 3.13 obtained by extending one in Fig. 3.9 is the smallest solution. There is only one orthogonal polyhedron with a surface area of 10 shown in Fig. 3.9, and there are only two orthogonal polyhedra with a surface area of 14 shown in Fig. 3.13. Therefore, the net that can fold (in three ways) is a solution to the smallest area.

Then, there is a common net that can fold any number of orthogonal polyhedra? We give an affirmative answer to this natural question below.

Theorem 3.3.3 *For any positive integer k, there is a net that can fold into at least k different orthogonal polyhedra.*

Proof Here we use a common net that can fold into two boxes of size $1 \times 1 \times 8$ and $1 \times 2 \times 5$ shown in Fig. 3.14a as a gadget. We call the black squares in Fig. 3.14a

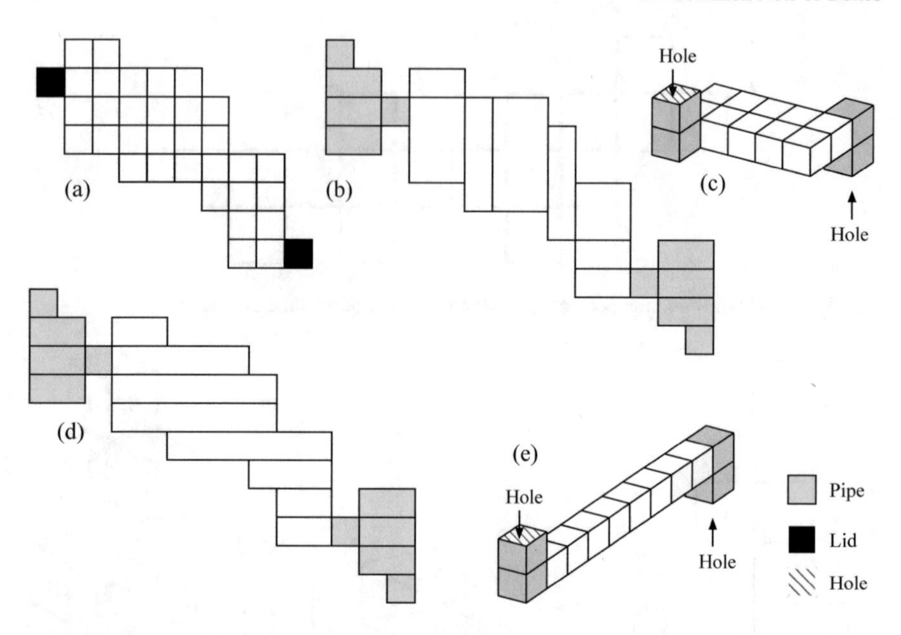

Fig. 3.14 Connection gadget

“lids”. They are placed on the opposite side in either folding of the two boxes. The gadget shown in gray in Fig. 3.14b–e is called “pipe”. If you replace the two lids with pipes, you can make two “holes” on the opposite side as shown in Fig. 3.14c, e in either box. We prepare k' of this gadget and join together. Then, we leave two lids at both ends as is without replacing by the pipes. An example of $k' = 3$ is shown in Fig. 3.15.

Since each pair of pipes (or lids) will be placed on the opposite side in either box that is folded, each net can be independently folded into any box of size either $1 \times 1 \times 8$ or $1 \times 2 \times 5$. Therefore, we can fold $2^{k'}$ polyhedra in total of which $2^{\lceil k'/2 \rceil}$ are inversion symmetry. Thus we can fold $(2^{k'} - 2^{\lceil k'/2 \rceil})/2 + 2^{\lceil k'/2 \rceil} = 2^{k'-1} + 2^{\lceil k'/2 \rceil - 1}$ different orthogonal polyhedra. (For example, the net in Fig. 3.15 can be folded into $4 + 2 = 6$ different orthogonal polyhedra.) Hence, by taking a sufficiently large k', we have the theorem. □

3.4 Common Nets of Three Boxes

In Sect. 3.2, we showed various algorithms and results on the nets in which two boxes can be folded. In the series of results, the question as to whether there exist common nets of three or more boxes arises very naturally. Particularly noteworthy, one is the net in Fig. 3.16, which was found in the 2263 common nets obtained through research in Sect. 3.2.3. This net can fold into not only two boxes of size $1 \times 1 \times 5$

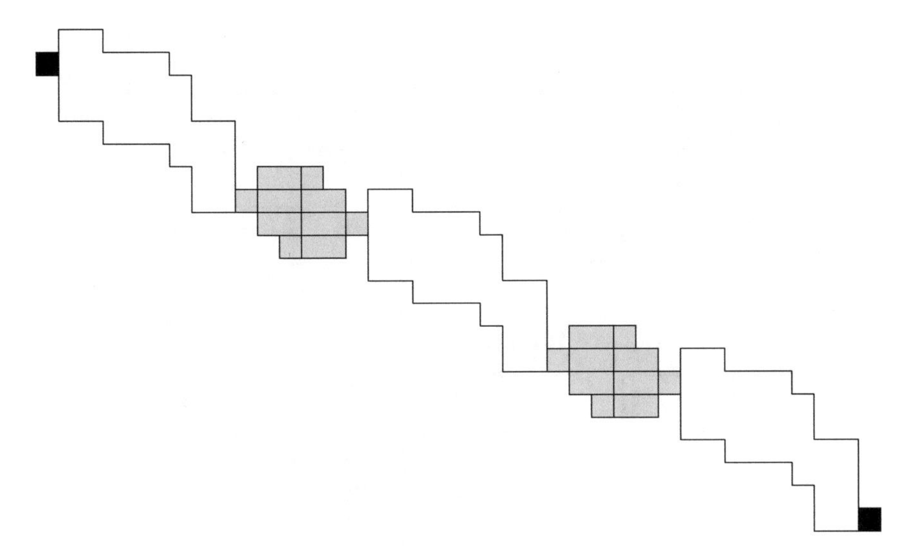

Fig. 3.15 A net by joining three gadgets together

and $1 \times 2 \times 3$, but also a "box" of size $0 \times 1 \times 11$. The last one is a doubly covered rectangle, which is a degenerated box. There is a difficulty comparable to something of a puzzle to fold the other two boxes, but if you look closely at the figure, you can understand that the doubly covered rectangle can be folded. Since two squares are arranged in each column vertically, you can round it vertically to obtain a cylinder. You still have extra squares at each end, then you can fold it in half to close the cylinder. Note that there is only one of 2263 nets with this property. (This special net was found "visually" by Mr. Hiroaki Matsui who belonged to my laboratory in 2010. Although it can be convinced if you understand to a certain extent, it is my hat's off to his enthusiasm which coincidentally found this, which is the only one among 2263 nets.)

Anyway, thanks to this example, it seems that there may be slightly more "certain" common nets of three different boxes. In this section, we show that there are infinitely many common nets of three boxes. Before that, we will point out that there are two strategies to search such nets.

The first one is a simple method of extending the search space. However, this is not so easy. As the area increases, the search space exponentially increases. As an example, consider area 22 and area 88. By subdividing the unit square into four squares, we can create a solution of $P(44)$ from a solution of $P(11)$. That is, among the sets $P(11) = \{(1, 1, 5), (1, 2, 3)\}$ and $P(44) = \{(1, 2, 14), (1, 4, 8), (2, 2, 10), (2, 4, 6)\}$, $(1, 1, 5)$, and $(1, 2, 3)$ correspond to $(2, 2, 10)$ and $(2, 4, 6)$, respectively. Observing these solutions, you may think that if you apply the first algorithm shown in Sect. 3.2.1 to $P(44)$ as it is, the algorithm will manage somehow. In fact, the first algorithm randomly generated 6.7×10^7 trials (in 3 days) for $P(11)$ and output 541 solutions, on the other hand, only two solutions were obtained for $P(44)$, although random

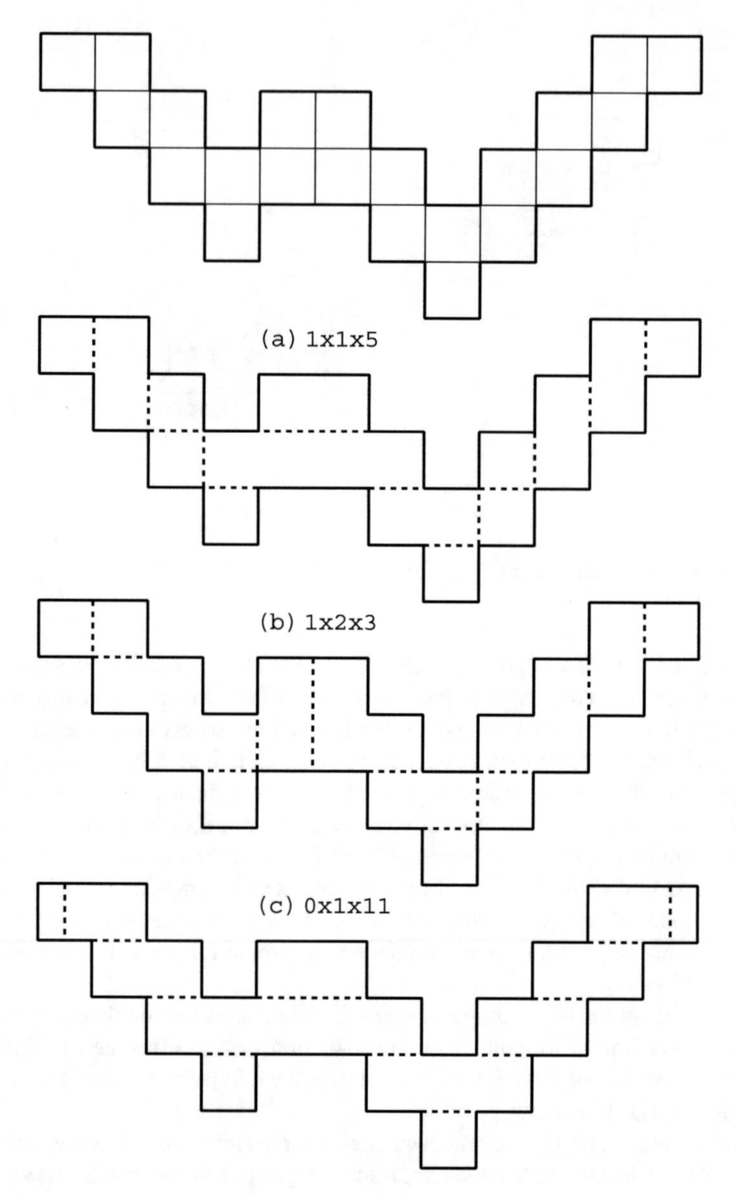

Fig. 3.16 A common net of not only two boxes of size $1 \times 1 \times 5$ and $1 \times 2 \times 3$, but also a “box” of size $0 \times 1 \times 11$

generation of 217.0×10^7 trials (over about a month) was done. Actually, the first algorithm works for a certain area of S, so it ran for various areas for several months. When we examine the results in Table 3.1 carefully, for example, even if it is of area 46 or 54, we have searched for only small amount in the search space, and hence it is very unlikely to find out by chance. (The enumeration problem for polyominoes seems to be much simpler than the enumeration problem for nets. However, as of 2018, the number of the i-ominoes is only known up to $i = 45$. Therefore, even with the strategy of finding out all the nets of the area i, the smaller size seems to be the limit.) Actually, running a program on a supercomputer for several months did not yield any common net that can fold into three (or more) different boxes. Thus, some ingenuity is required.

The other strategy is to find the common nets of three boxes using some properties of nets and new ideas without relying on search. We will describe successful methods and results of these two strategies.

3.4.1 *Search for Special Area 30*

As mentioned above, trying to fold three or more boxes in a normal way requires an area of at least 46. In case of $P(23) = \{(1, 1, 11), (1, 2, 7), (1, 3, 5)\}$, we have to search common nets of three boxes; however, area 46 is too big to search. Now we consider the area 30, which corresponds to $P(15) = \{(1, 1, 7), (1, 3, 3)\}$. The area 30 is the next candidate for area 22, i.e., $P(11) = \{(1, 1, 5), (1, 2, 3)\}$. Although 3-tuples of natural numbers (a, b, c) that satisfy $2(ab + bc + ca) = 30$ for a box with a surface area of 30 are only $(1, 1, 7)$ and $(1, 3, 3)$, we change out a way of thinking here. For example, what if $a = b = c$? Then we have $6a^2 = 30$, and $a = b = c = \sqrt{5}$ is obtained. Now $\sqrt{5}$ is the length of the diagonal of a rectangle of size 1×2. Therefore, if we allow to fold diagonally, we may have a common net of not only two boxes of size $1 \times 1 \times 7$ and $1 \times 3 \times 3$, but also a cube of size $\sqrt{5} \times \sqrt{5} \times \sqrt{5}$. (Incidentally, this fact was first noticed by Mr. Toshihiro Shirakawa, who is my puzzle friend. This bold idea should be admired.) Then, we introduce the results of the search using the following algorithm.

1. Find the set D of common nets of boxes of size $1 \times 1 \times 7$ and $1 \times 3 \times 3$.
2. For each element in D, check whether it can fold into the cube of size $\sqrt{5} \times \sqrt{5} \times \sqrt{5}$.

If breadth-first search of the nets of area 22 could be done in about 5 h on a PC as of 2015, you may think that breadth-first search of the nets of area 30 would be somehow manageable. However, this cannot be done straightforwardly. Using the supercomputer (CRAY XC 30) at my workplace, when trying to examine all common nets of the boxes of size $1 \times 1 \times 7$ and $1 \times 3 \times 3$ by breadth-first search, the memory overflowed at the area 22. There are too many common partial nets. Therefore, in the ordinary way (or natural extension of the algorithm for area 22), it is not possible to enumerate all the common nets of boxes of size $1 \times 1 \times 7$ and $1 \times 3 \times 3$. Thus, further ingenuity is required. My research group solved this problem in two different ways, but since it goes beyond the scope of this book, we will skip the details and let us briefly outline the method (the details can be found in [XHS+17]).

Before the details, we here state the conclusion: There are 1080 common nets of two boxes of size $1 \times 1 \times 7$ and $1 \times 3 \times 3$. Among them, there are nine nets that can fold into the cube of size $\sqrt{5} \times \sqrt{5} \times \sqrt{5}$ (Fig. 3.17). To my surprise, among these nine nets, two nets are really interesting; there are four ways to fold these three boxes. One net has two ways to fold into the cube of size $\sqrt{5} \times \sqrt{5} \times \sqrt{5}$ (Fig. 3.18), and another net has two ways to fold into the box of size $1 \times 3 \times 3$ (Fig. 3.19). The results that there are 1080 common nets of boxes of size $1 \times 1 \times 7$ and $1 \times 3 \times 3$, and that nine of them can fold into the cube were within the expected range. However, I did not expect that there would be the nets that have four different ways to fold three boxes as in Figs. 3.18 and 3.19. It can be said that these nets are absolute gems.

3.4.1.1 How to Modify Brute Force Algorithm

Even in the case that the memory overflows halfway in a simple breadth-first search, a method of conserving memory by combining breadth-first search and depth-first search is conceivable. Specifically, in my research group, we first performed breadth-first search up to area 16. As a result, 7486799 common partial nets were obtained in the area. Next, we divided them into 75 groups, performed the depth-first search for each group, and separately obtained common nets of area 30. Finally, we merged these to obtain all common nets. Considering the computation resources of the supercomputer in my workplace as of 2015, the memory and the calculation time were balanced. In this way, using a supercomputer (CRAY XC 30), the computation ended successfully less than 3 months in total.

3.4.1.2 How to Modify an Algorithm

Here we introduce some ideas on algorithm and data structure. The algorithm itself is close to a method of generating a spanning tree. Specifically, we adopted the following method.

1. Consider "cutting" and "not cutting" for each edge of unit squares in the box of size $1 \times 1 \times 7$, and memorize only those which are nets in all combinations.
2. Consider "cutting" and "not cutting" for each edge of unit squares in the box of size $1 \times 3 \times 3$, and memorize only those which are nets in all combinations.
3. Extract only the common ones in the above two nets.

Implementing this with a simple data structure, memory will overflow on the way. In recent years, what is called a *binary decision diagram (BDD)* has emerged as a data structure that can efficiently operate such data. We managed nets using BDDs and developed a program to take these common parts at the end. With this method, considerable speeding up can be realized, and finally it has become possible to

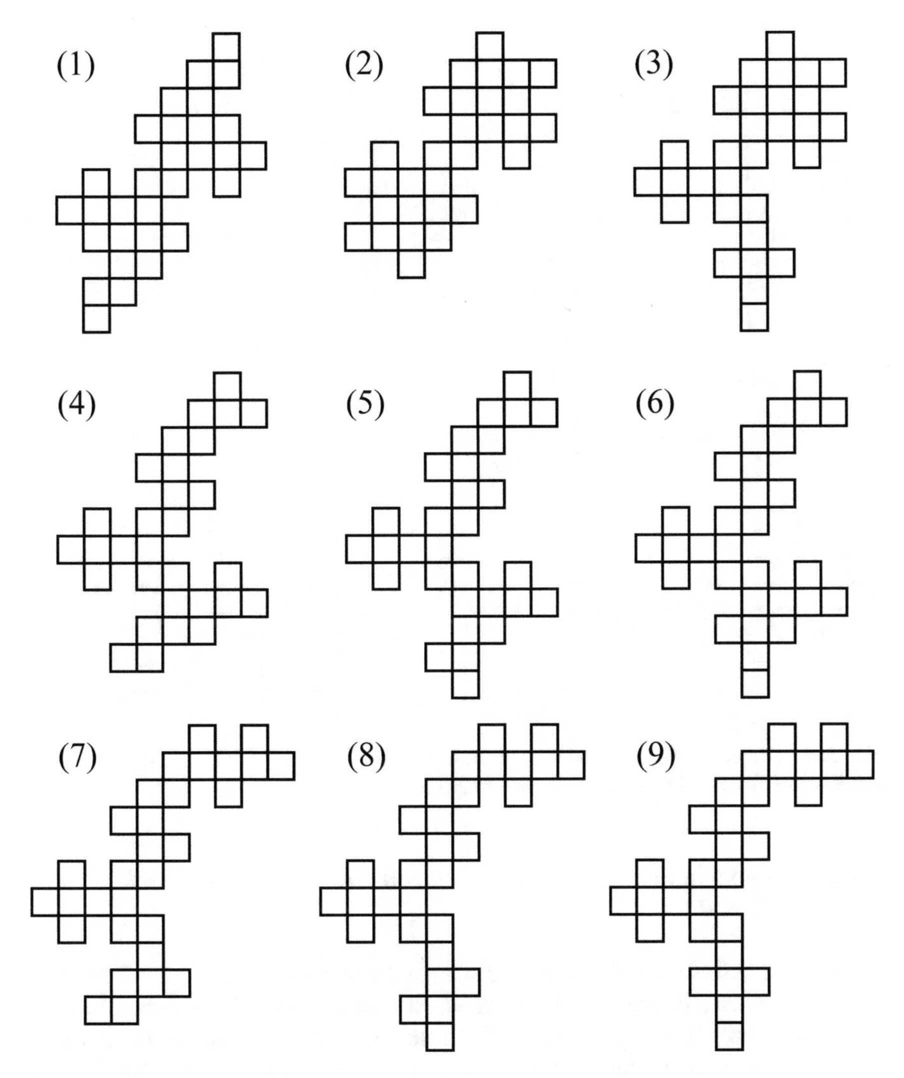

Fig. 3.17 Nine polyominoes of area 30 that can fold into two boxes of size $1 \times 1 \times 7$ and $1 \times 3 \times 3$, and one cube of size $\sqrt{5} \times \sqrt{5} \times \sqrt{5}$

compute with a high-spec computer of 3.3 GHz with 128 GB of memory mounted in less than 10 days.

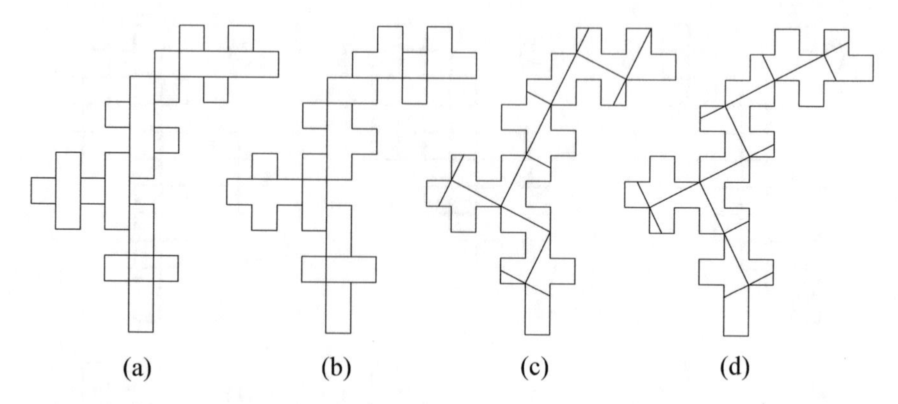

Fig. 3.18 Unique net that can fold into two boxes of size $1 \times 1 \times 7$ and $1 \times 3 \times 3$, and one cube of size $\sqrt{5} \times \sqrt{5} \times \sqrt{5}$ in two different ways

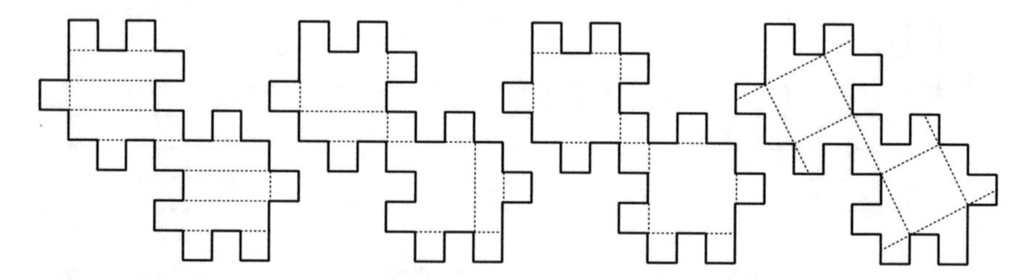

Fig. 3.19 Unique net that can fold into a cube of size $\sqrt{5} \times \sqrt{5} \times \sqrt{5}$, a box of size $1 \times 1 \times 7$, and a box of size $1 \times 3 \times 3$ in two different ways

3.4.1.3 Algorithm for Checking if a Polygon Folds to a Cube of Size $\sqrt{5} \times \sqrt{5} \times \sqrt{5}$

We found that there are 1080 common nets of the boxes of size $1 \times 1 \times 7$ and size $1 \times 3 \times 3$. Then it is necessary to check whether the cube of $\sqrt{5} \times \sqrt{5} \times \sqrt{5}$ can be folded from each of them. Generally, determining whether the polyhedron Q can be folded from a given polygon P is not an easy problem. Recently, my colleagues and I give an efficient algorithm in [MHU19] for solving this folding problem for any given polygon P and a box Q. Here, we explain a simpler algorithm specialized for the problem by utilizing that Q is a specific cube and P is a polyomino of 30 unit squares, which was used in [XHS+17, Xu17]. We will introduce a solution to this problem with some general issues below.

The polyhedron Q here is a cube of size $\sqrt{5} \times \sqrt{5} \times \sqrt{5}$, and the polygon P is a polyomino in which 30 unit squares are connected. Therefore, the crease lines should be as in Fig. 3.20 (or its mirror image) if P can be folded into Q. Here, by Theorem 2.1.3, each vertex of Q is a point on the boundary of P. Moreover, as you can see in Fig. 3.20, each vertex of Q is actually a grid point on P. That is, there is a possibility that only grid points on the boundary of P will be vertices of the cube Q. Conversely, on the cube Q, there are only two cases for grid points on the boundary

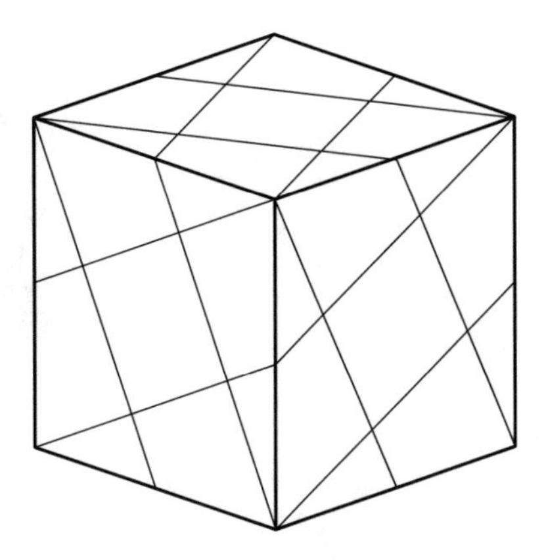

Fig. 3.20 The appearance of the cube of size $\sqrt{5} \times \sqrt{5} \times \sqrt{5}$ folded from polyomino of area 30

of P; each of them becomes either a vertex of Q or a vertex of square corners at the center of each face of Q.

Specifically, we explain the algorithm using polyomino at the upper left of Fig. 3.17. There are only 10 ways of overlaying the grid of P on this polyomino as shown in Fig. 3.21. Looking at the top gray square, each of the four corners of this square may be grid points, but considering each mirror image, there are eight ways. There are two more cases where this gray square becomes a square in the center of the face of Q. Thus, there are 10 ways in total. Of these, grid points are inside of the polyomino (indicated by $\circ$ in the figure) in (1), (3), (4), (5), (7), (8), and (10). Since these grid points correspond to a vertex of Q, four squares should not gather here. Therefore, they can be excluded from the candidates.

In cases (2), (6), (9) where all grid points are on the boundary, we need to make sure properly. An algorithm for determining whether or not a given net can fold into a convex polyhedron in such a manner is published in [DO07]. The algorithm given in [DO07] essentially tries all pairs of edges to be glued and outputs all convex polyhedra if possible. For our problem, we modify this algorithm slightly to check if the cube can be folded as follows.

1. At the grid points on the boundary of the polygon P, glue the corresponding edges where three squares are gathered (indicated by $\circ$ in Fig. 3.22).
2. After that, begin with each glued part and decide pairs of edges to be glued. At this time, by the shape of Q, the number of squares gathered at each point can be fixed to three if it is a grid point, otherwise to four.
3. If unit squares overlap each other halfway, the cube cannot be folded. Otherwise, it can be folded.

Unlike the algorithm of literature [DO07], this algorithm runs in linear time because the number of unit squares surrounding each grid point is fixed.

Examples of the execution are shown in Fig. 3.22(2), (9). In the case of (2), if gluing is extended starting from the gluing of the point with $\circ$, unit squares with $\times$ overlap each other, and it fails to fold the cube. On the other hand, in the case of

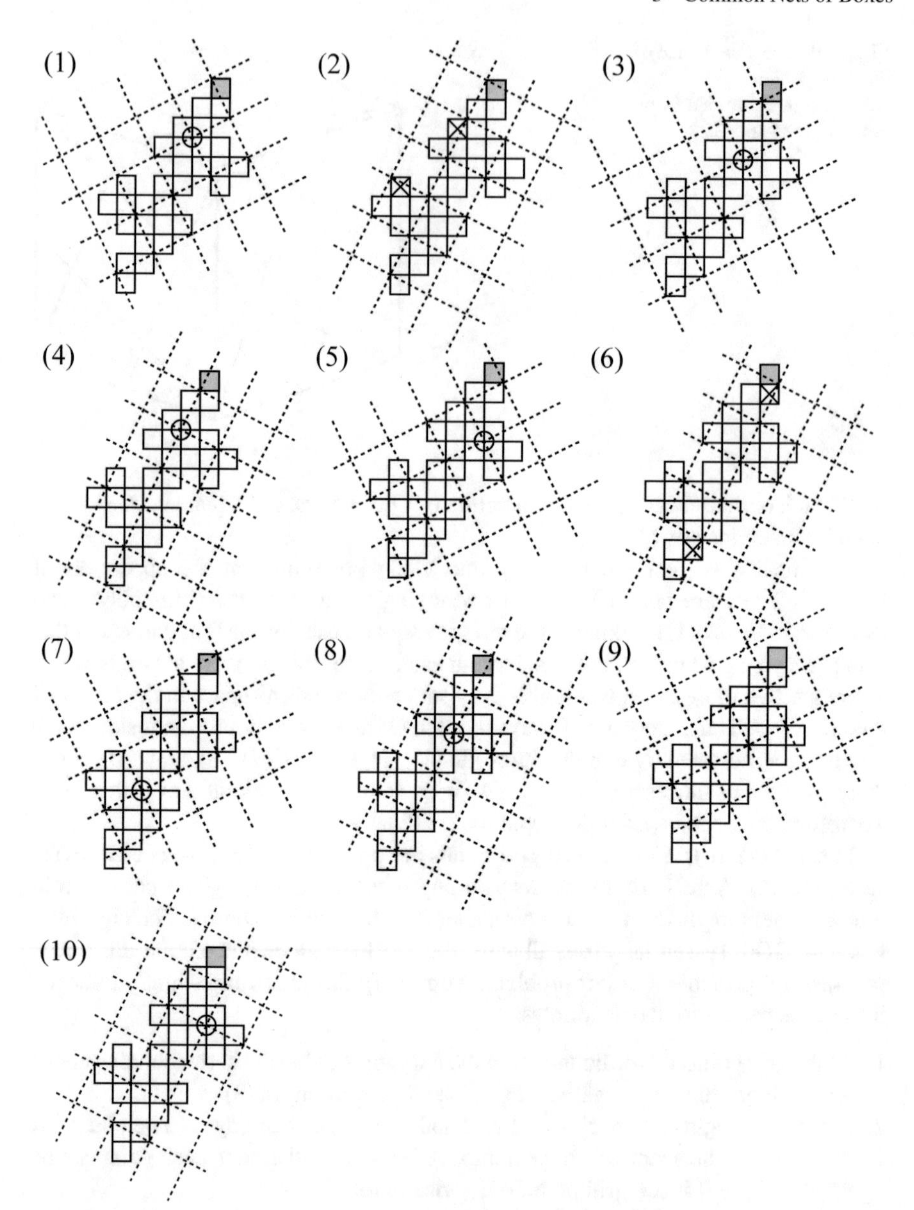

Fig. 3.21 A method of checking whether a polyomino can fold into a cube (1)

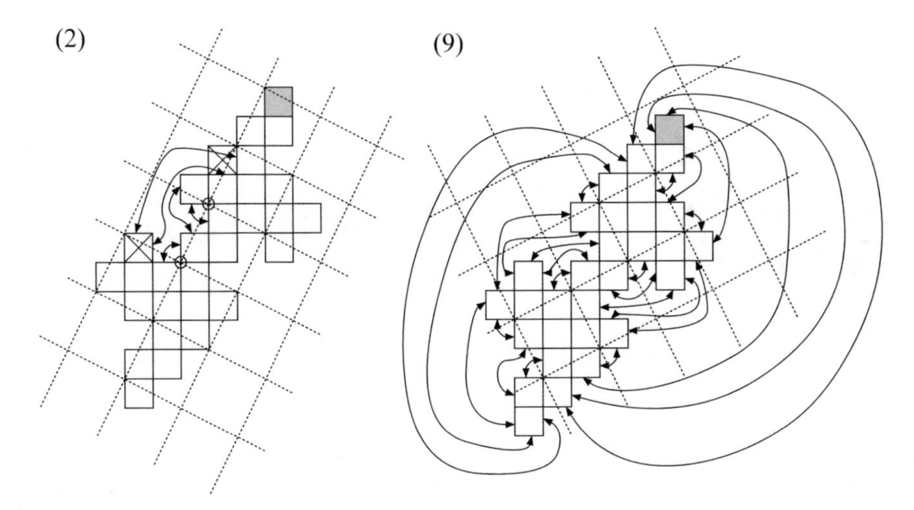

Fig. 3.22 A method of checking whether a polyomino can fold into a cube (2)

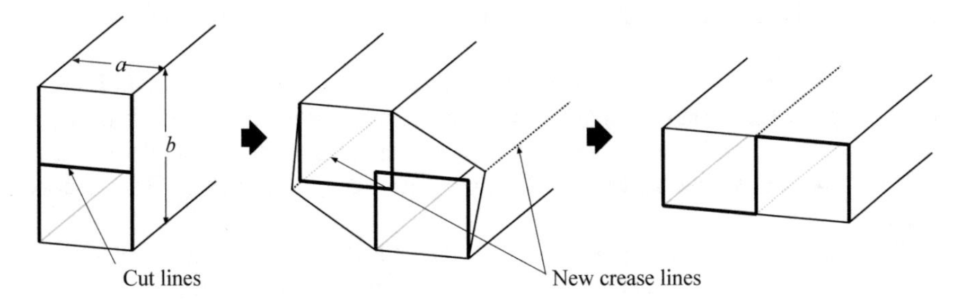

Fig. 3.23 Basic idea: "crush" box

(9), when extending gluing starting from the point with ∘, the whole is completed without contradiction, and the cube can be folded with these crease lines.

However, with such a big and complicated net, even if the program outputs an affirmative answer, it is difficult to imagine. For example, I want you to try out the net in Fig. 3.18 by all means.

Exercise 3.4.1 *Fold the net of Fig. 3.18 into the boxes in practice.*

3.4.2 Construction Based on a New Idea

Next, I introduce a method for constructing "common nets of three boxes" based on an entirely new idea. The basic idea is a "modification" of a common net of two boxes to fold the third box.

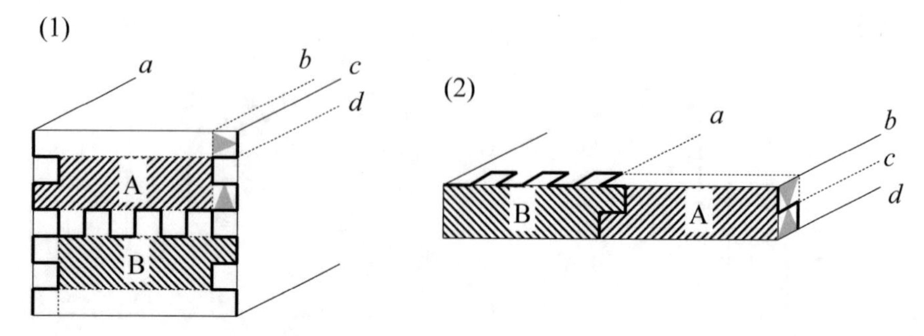

Fig. 3.24 Cut lines and crease lines for crushing the box

First, prepare a net that can be folded into two boxes of size $a \times b \times c$ and $a' \times b' \times c'$. Pay attention to two rectangles of size $a \times b$ of box of size $a \times b \times c$. We call them a *lid* and a *bottom*. Make incisions in the "H" shape at the lid and bottom, and divide them into two halves of size $a \times b/2$. Then you crush the box as in Fig. 3.23 and create a box of new size. Both the lid and the bottom have a size of $(a + b/2) \times b/2$.

However, this idea does not work as it is. The size of the lid and the bottom is $(a + b/2) \times b/2$, but at this time, in order for a rectangle to be properly formed, it must be $a = b/2$ or $2a = b$. Then this is only transforming the rectangle of 1×2 to 2×1, so it does not become a "different" box (this is essentially the same as Fig. 3.9).

The main idea for avoiding this issue is to move a part of the lid and the bottom to the side so as to change the area of the lid and bottom after deformation. The person who first came up with this idea was Mr. Toshihiro Shirakawa, one of my puzzle friends. Although the idea of "crushing a box" was pointed out by other researchers, nothing comes to mind such a novel idea of changing the area. A concrete cutting pattern is shown in Fig. 3.24. The bold solid lines are the cutting lines, and a to d are the corresponding crease lines. In the figure, the parts A and B with slanting lines are the bottom even after crushing in (2), and the other surrounding parts wrap around the side of the box. Looking at the gray triangles, you will see how this trick works. In this example, the rectangle of size 8×7 becomes the rectangle of size 13×2 after the bottom is cut with zigzag lines and crushed the box. The area changes from 56 to 26, and the difference 30 moves to the sides of the box. We also note that the perimeter of the two rectangles does not change as $7 + 8 + 7 + 8 = 2 + 13 + 2 + 13 = 30$.

In order to make use of this idea, we have to choose a net with a nice property. The net chosen by us is shown in Fig. 3.25. This is a common net of two boxes of size $a \times b \times 8a$ and $a \times 2a \times (2a + 3b)$. Actually, this net was originally chosen from common nets of the boxes of size $1 \times 1 \times 8$ and $1 \times 2 \times 5$. To successfully implement the idea, we divided the lid and bottom of the size $a \times b$ into two rectangles of size $a/2 \times b$. Originally, it is a net based on the unit square of $a = b$, but in the case of this net, you can choose any positive real numbers of a and b independently.

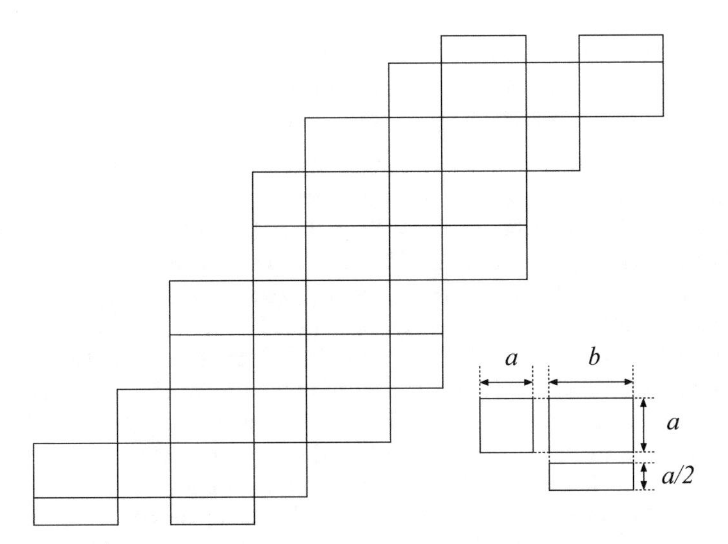

Fig. 3.25 The base common net of two boxes. For arbitrary positive real numbers a and b, one is of size $a \times b \times 8a$, and the other is of size $a \times 2a \times (2a + 3b)$

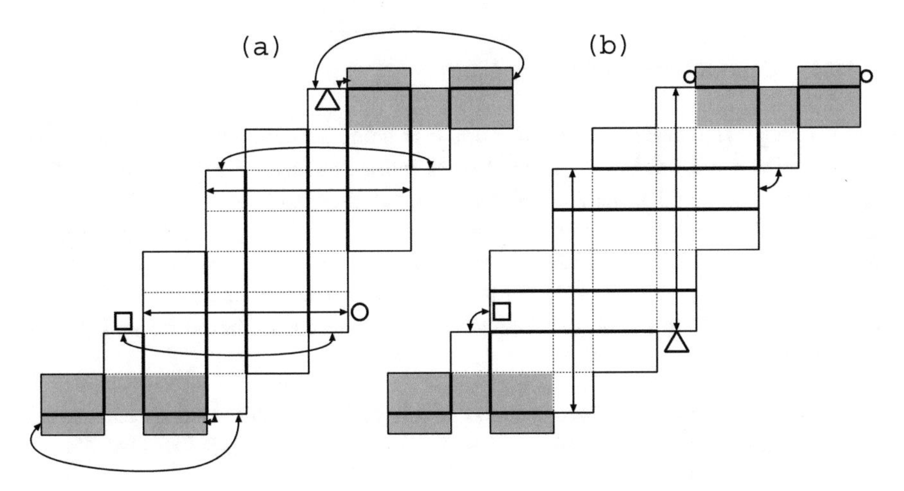

Fig. 3.26 The gluing correspondence for folding to two boxes of size $a \times b \times 8a$ and $a \times 2a \times (2a + 3b)$

If you look carefully at the bold lines in Fig. 3.26, you will see how to fold each box, and a and b are independent.

The net in Fig. 3.25 has good properties that can implement the idea of Fig. 3.23. In particular,

- a and b are independent, and each can be of any length. (In other words, even with this net, we can prove Theorem 3.3.1 in Sect. 3.3.4 that states there are infinitely many nets that can fold into two boxes. We here note that both a and b can be arbitrary real numbers, not only natural numbers.)

- In two ways of folding, some crease lines are common.

The latter property is important. If we can embed the zigzag lines of Fig. 3.24 in this common crease line part, we should safely fold the third box. The bold lines shown in Fig. 3.26 are common crease lines. Therefore, with $a = 7$ and $b = 8$, try embedding the zigzag lines of Fig. 3.24 in this common part. Then, a new problem arises here. In one of the two ways of folding, some zigzag lines will be glued to other places. For example, the zigzag lines at the bottom of Fig. 3.26a is glued to the $\triangle$ of Fig. 3.26b. Thus we also have to embed zigzag lines here. However, if the zigzag lines embedded in this part are folded in the way of Fig. 3.26a, they will be glued to the place of $\square$ in Fig. 3.26a. Therefore, we also have to embed zigzag lines here. In this way, we need to carefully follow this "spread of zigzag lines" with both folding ways on this net. Following this in order, the ripple effect revolves around and comes back. If the parity of zigzag mismatches here, this construction will fail; however, in the case of Fig. 3.25, it works well, and we finally get the common net of three boxes of size $7 \times 8 \times 56$, $7 \times 14 \times 38$, and $2 \times 13 \times 58$ (Fig. 3.27)!

3.4.2.1 Generalization of Zigzag Lines

In the previous example, we let $a = 7$ and $b = 8$ and deformed the rectangle of size 7×8 to size 2×13. It is not difficult to generalize this method. For example, if you set $a = 11$ and $b = 10$, the size of the rectangle changes from 11×10 to 4×17 (Fig. 3.28). In general, if you set $a = 4k + 7$ and $b = 2(k + 4)$ for arbitrary natural number $k = 0, 1, 2, \ldots$, in the same way of Fig. 3.24, you can transform the size of a rectangle from $a \times b$ to $2(k + 1) \times (4k + 13)$. The difference from Fig. 3.27 is only the number of the mountains and valleys of the zigzag lines. Therefore, we have the following theorem.

Theorem 3.4.2 *For any natural number* $k = 0, 1, 2, \ldots$, *there exists a common net of three different boxes of size* $(4k + 7) \times 2(k + 4) \times 8(4k + 7)$, $(4k + 7) \times 2(4k + 7) \times 2(7k + 19)$, *and* $2(k + 1) \times (4k + 13) \times 2(16k + 29)$.

In other words, it turned out that there are infinitely many common nets of three different boxes.

3.5 Summary of This Chapter and Open Problems

In this chapter, we introduced nets from various perspectives in which multiple boxes can be folded. It is conceivable that common nets of two boxes are likely to exist generally if two boxes are equal in area. We can say that research on the nets of three or more boxes is still on the way at present. There are obviously quite huge ones that

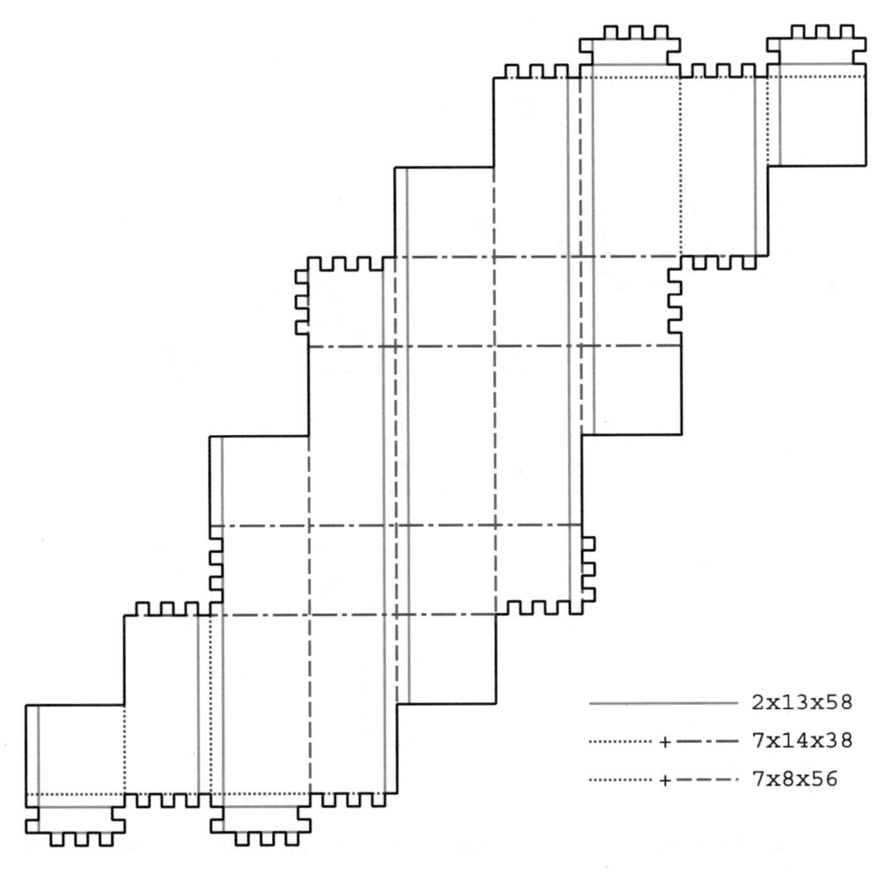

Fig. 3.27 Common net of three boxes of size $7 \times 8 \times 56$, $7 \times 14 \times 38$, and $2 \times 13 \times 58$

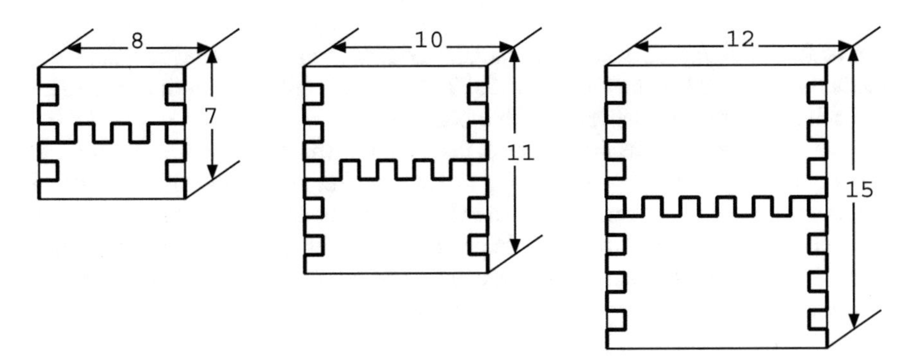

Fig. 3.28 Generalization of zigzag lines

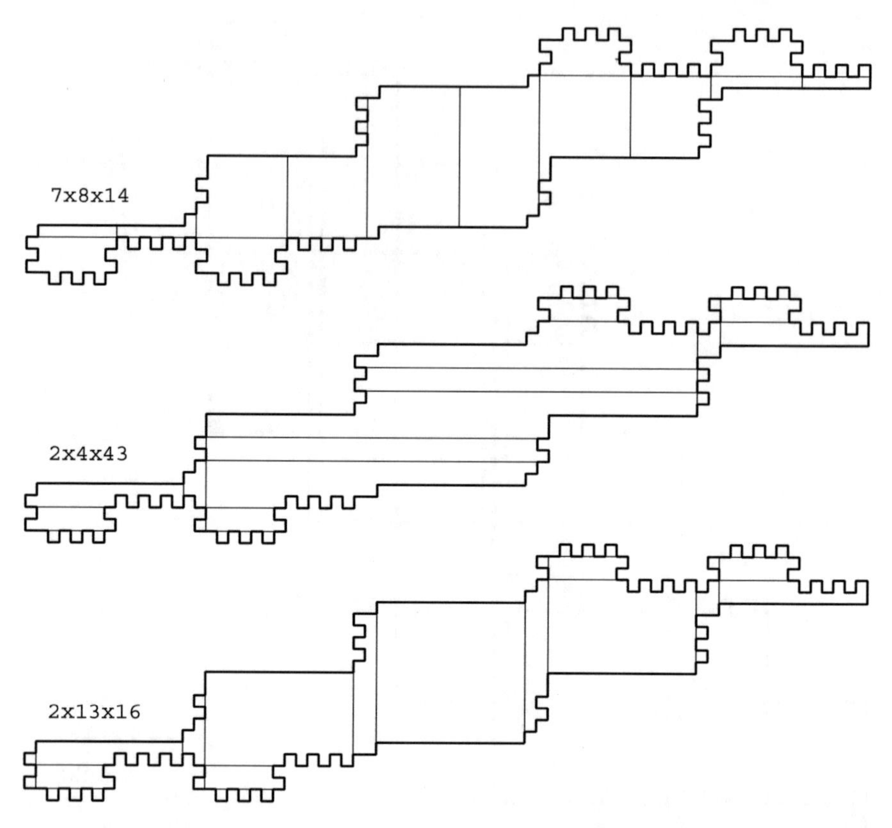

Fig. 3.29 Another common net of three different boxes. The area is 532. The size of boxes is $7 \times 8 \times 14$, $2 \times 4 \times 43$, and $2 \times 13 \times 16$

can be folded along the grid at this stage. A relatively small net constructed by the method of Fig. 3.29 is shown in Sect. 3.4.2. The area is still 532, but it seems to be difficult to find much smaller net with this method.

One of the most interesting issues is the area 46, where three boxes of size $1 \times 1 \times 11$, $1 \times 2 \times 7$, and $1 \times 3 \times 5$ might be folded from one polygon. It is certain that the computational resources overflow if the current area 30 is extended as it is, and hence it is necessary to further improve the algorithm considerably. The most interesting problem is whether there are nets in which four or more different boxes can be folded. Of course, no one has been found yet, and we do not know how to find it at the moment.

It is quite unlikely that one polygon can fold into thousands of different boxes; however, how can we show the upper bound of the number of possible boxes to be folded from one polygon? (Remind that Theorem 3.1.1 states that there is no upper bound of this number from the viewpoint of the surface area.)

3.5.1 Rotationally Symmetric Nets

When you look at the nets in this chapter, you can see some impressive nets from time to time. Especially, it seems that there is something meaningful in the nets with high symmetry. Consider rotationally symmetric nets such as Figs. 3.5(1), (3), (8), 3.7, and 3.17(1), (2). It has several advantages to tackle the above open problems limited to such rotationally symmetrical nets. First, since it is enough to memorize only half of the net, the amount of memory can be considerably reduced. Also, the area can be increased by 2 instead of 1, so it is expected that the computation time can be considerably shortened. In fact, Mr. Dawei Xu, a Ph.D. student in my laboratory, explored symmetric nets in 2017. However, the results were somewhat disappointing [Xu17]:

Theorem 3.5.1 *Regarding the area 46 and the area 54, when it is restricted to rotationally symmetrical nets, although there exist common nets of two boxes (for each pair), there is no rotationally symmetrical common net of three boxes.*

However, if it comes to an area 70, for example, since it is consistent with four boxes of size $1 \times 1 \times 17$, $1 \times 2 \times 11$, $1 \times 3 \times 8$, and $1 \times 5 \times 5$, there may exist a rotationally symmetrical common net of three of these four boxes. Research and inherent theorem on such characteristic nets are also interesting themes.

3.6 Extra Problem

Here, we stray somewhat from the main topic, but let us introduce one example that you will fall into a pitfall if you consider a net or a polyhedron with just intuition. In this chapter, we investigated "boxes" which can be folded with polygon on square grid mainly because it is easy to handle with computer. By the way, when "a polyhedron whose faces are all rectangles" is given, is it true that the whole angles made by two faces are always multiples of 90°? At first glance, you may think quite naturally that it is [Yes]. However, it is a surprising fact that although all the faces are rectangular, there is a polyhedron that the whole angles between two faces are not multiples of 90°. Moreover, it is not so difficult to understand once you see it. How about trying this as brain teaser?

Exercise 3.6.1 *What is a polyhedron whose angles between two faces are not multiples of* 90°, *although all faces are rectangular?*

Consider carefully before seeing the answer. If you make this real solid with paper, it will be pretty fun.

References

[AT08] T. Asano, H. Tanaka, Constant-working space algorithm for connected components labeling, in *IEICE Technical Report*, vol. COMP2008-1 (2008), pp. 1–8

[BCD+99] T. Biedl, T. Chan, E. Demaine, M. Demaine, A. Lubiw, J.I. Munro, J. Shallit, Notes from the University of Waterloo algorithmic problem session, 8 September 1999

[DHK+19] M.L. Demaine, R. Hearn, J.S. Ku, R. Uehara, Rectangular unfoldings of polycubes, in *Canadian Conference on Computational Geometry* (2019), pp. 164–168

[DO07] E.D. Demaine, J. O'Rourke, *Geometric Folding Algorithms: Linkages, Origami, Polyhedra* (Cambridge University Press, Cambridge, 2007)

[Gar08] M. Gardner, *Hexaflexagons, Probability Paradoxes, and the Tower of Hanoi: Martin Gardner's First Book of Mathematical Puzzles and Games* (Cambridge University Press, Cambridge, 2008)

[Gol94] S.W. Golomb, *Polyominoes* (Princeton University Press, Princeton, 1994)

[KRU07] M. Kano, M.-J.P. Ruiz, J. Urrutia, Jin Akiyama: a friend and his mathematics. Graphs Comb. **23**(Suppl), 1–39 (2007)

[MK06] G. Miller (D.E. Knuth), Cubigami, *Cubism for Fun*, vol. 70 (2006), pp. 24–27, http://www.puzzlepalace.com/

[MHU19] K. Mizunashi, T. Horiyama, R. Uehara, Efficient algorithm for box folding, in *International Conference and Workshop on Algorithms and Computation (WALCOM 2019)*. Lecture Notes in Computer Science, vol. 11355 (2019), pp. 277–288

[Xu17] D. Xu, Research on developments of polycubes, Ph.D. thesis, Japan Advanced Institute of Science and Technology (2017)

[XHS+17] D. Xu, T. Horiyama, T. Shirakawa, R. Uehara, Common developments of three incongruent boxes of area 30. Comput. Geom.: Theory Appl. **64**, 1–17 (2017). https://doi.org/10.1016/j.comgeo.2017.03.001

Chapter 4
Common Nets of (Regular) Polyhedra

Abstract In Chap. 3, we consider box, which is one of the most natural shapes in our daily life. From the viewpoint of mathematics, we have some other natural shapes. For example, we have five regular polyhedra which have been investigated since Archimedean's period. However, surprisingly, they are not yet well investigated from the viewpoint of the unfolding. For example, it is not known that two regular polygons have a common net. That is, we do not know whether there is a polygon that can fold to two or more regular polyhedra. We give positive and negative results related to this interesting open problem.

4.1 Classification of Regular Polyhedra

For unfoldings, intuition often does not work. Even though it seems easy at first glance, you may notice something that is not easy after you give considerable thoughts. As you can see in Fig. 2.8, generally, various polyhedra can be folded from one polygon. There is little mathematical knowledge about the relationship between these polygons and the polyhedra that can be folded from them. Introducing the problem in a general way, it can be shown as follows.

Open Problem 4.1.1 *For a given polygon P and a polyhedron Q with the same surface area as P, determine whether Q can be folded from P or not.*

For this problem, there is no clue in any way to solve it in a general form at present. It is the current situation that this problem can be solved only in limited cases. In connection with this problem, the following open problem is famous.

Open Problem 4.1.2 *Is there a common net of multiple regular polyhedra? Namely, is there a polygon that can fold into two (or more) regular polyhedra?*

As shown in Sect. 2.2, there are only five types of regular polyhedra. They are also called Platonic solids that kept people's attention from the era of Plato. The problem is unresolved even for such familiar solids. At first glance, this may seem "obviously impossible"; however, looking at the non-trivial nets in Fig. 2.8 or the splendid net of Fig. 4.1, we could feel like something possible. In this chapter, what we know about

R. Uehara, *Introduction to Computational Origami*,
https://doi.org/10.1007/978-981-15-4470-5_4

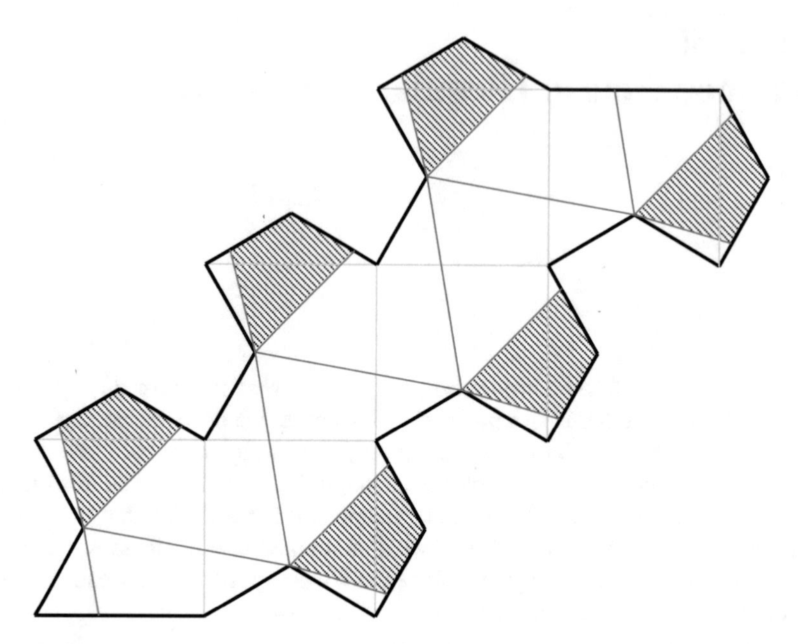

Fig. 4.1 A common net of a cube and an "almost" regular octahedron. A cube can be folded when folding along the vertical/horizontal gray lines in the figure. The octahedron can be folded by folding along diagonal lines. In this octahedron, the diagonally shaded parts in the figure are gathered three by three to form two regular triangular "lid" and "bottom", and their peripheries are surrounded by six isosceles triangles. This net that can only be described as wonderful was invented in 2010 by Mr. Toshihiro Shirakawa, who is one of my puzzle friends

nets for such familiar and beautiful three-dimensional polyhedra is introduced. Even in very limited cases, there is something you have to consider; on the other hand, you will find many interesting results that are counterintuitive. First, we introduce classification for several polyhedra, including regular ones.

4.1.1 Regular-Faced Convex Polyhedra

As mentioned above, there are only five types of regular polyhedra. However, when relaxing the condition a little, there are many convex polyhedra that are properly regulated. For example, a convex polyhedron such as a standard soccer ball that uses regular hexagons and regular pentagons in a well-balanced manner would be familiar. Since ancient times, "properly arranged convex polyhedra" have been well studied. A *regular-faced convex polyhedron* is a convex polyhedron that consists of regular polygonal faces such that all edges are equal in length. (Some readers may think that all edges are equal if all faces are regular polygons. When it is a natural polyhedron, that is correct. However, for example, considering an unnatural situation

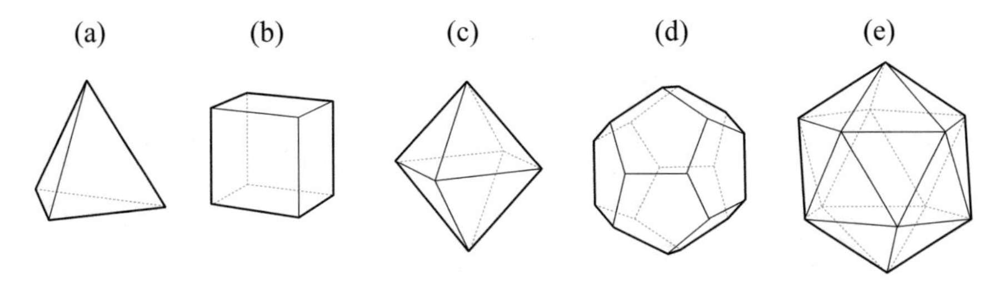

Fig. 4.2 Five regular polyhedra

such as a solid where it has a square face made by four edges of length 2, and two regular triangles are connected to one of these edges, the solid cannot be properly regulated anymore. Therefore, such seemingly redundant condition is added.) The class of regular-faced convex polyhedra is slightly wider than regular polyhedrons. Hereafter, let us assume that the unit length of an edge is 1 in this chapter. Then, the regular-faced convex polyhedra can be classified as follows.

Regular polyhedra: Convex polyhedra such that all faces are congruent regular polygons, and the number of polygons gathered at each vertex is also equal. As shown in Sect. 2.2, there are only five types of regular polyhedra. Specifically, they are a regular tetrahedron, a cube, a regular octahedron, a regular dodecahedron, and a regular icosahedron (Fig. 4.2a, b, c, d, e, respectively).

Semi-regular polyhedra: A *semi-regular polyhedron* is composed of two or more types of regular polygonal faces, and regular polygons gathered at each vertex are of the same kind and order. Specifically, there are 13 types in total. Without going into detail, 13 types are shown in Fig. 4.3. For example, the left end of the middle row is a familiar soccer ball shape, which is a representative semi-regular polyhedron and is called *truncated icosahedron*. The truncated icosahedron means "a solid which cut off the apexes of regular icosahedrons", and you will see its structure as you look closely at the figure. You can find that the soccer ball consists of 20 pieces of regular hexagons in this figure. Furthermore, since "the regular icosahedron and the regular dodecahedron are dual", it can be seen that there are 12 pentagons on it. Also, note that out of the 13 types, the last two are not identical to their own mirror image. Sometimes, we distinguish between them and their mirror images by calling "left-hand type" and "right-hand type".

Archimedean prism and antiprism: An *Archimedean prism* is the prism that consists of two "lid" and "bottom" of regular n-gons, and n squares joining the lid and the bottom. Figure 4.4a shows an example of a three-sided prism. An *Archimedean antiprism* is the polyhedron obtained by twisting a little the Archimedean prism and filling with regular triangles arranged side by side between the lid and the bottom. Figure 4.4b shows an example of the four-sided antiprism. Since both exist for arbitrary n of 3 or more, they exist infinitely. Here, note that the Archimedean four-sided prism is a cube, and the Archimedean three-sided antiprism is a regular octahedron.

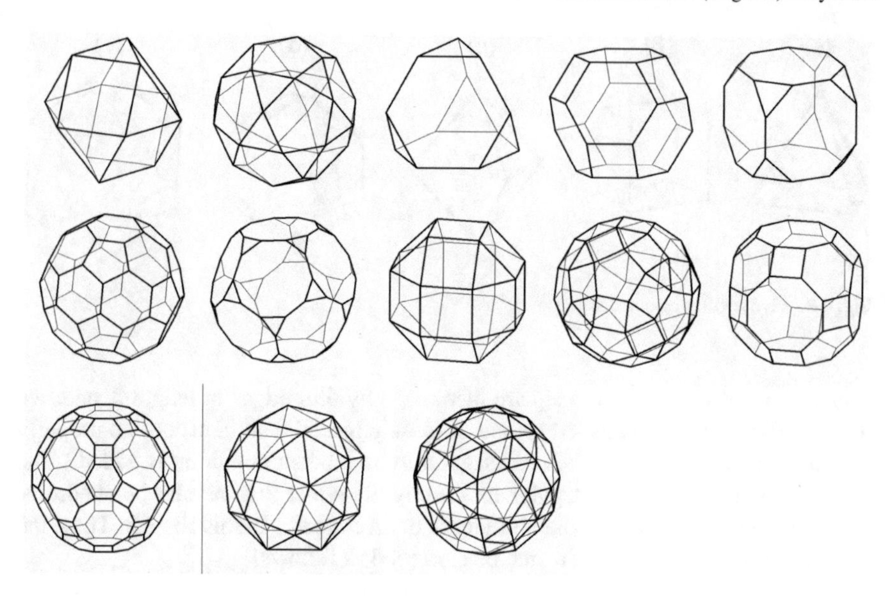

Fig. 4.3 Thirteen types of semi-regular polyhedra

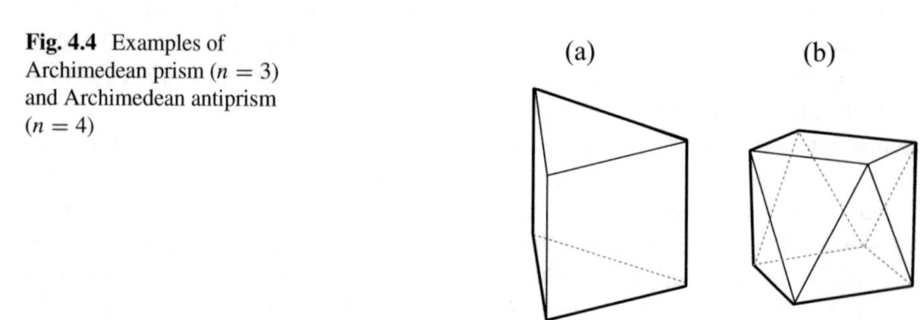

Fig. 4.4 Examples of Archimedean prism ($n = 3$) and Archimedean antiprism ($n = 4$)

Johnson–Zalgaller solid: Whereas enumeration of regular polyhedra and semi-regular polyhedra is not so difficult, classification of other convex polyhedra is not easy. Actually, an American mathematician Norman Johnson enumerated exhaustively, and Victor Abramovich Zalgaller confirmed it using a computer in 1969. There are 92 types of convex polyhedra excluding the regular polyhedra, the semi-regular polyhedra, the prisms, and the antiprisms. These polyhedra are numbered from J1 to J92 for convenience. In this book, we will refer to these 92 polyhedra as Johnson–Zalgaller solid. (They are sometimes called Johnson solids, and the entire regular-faced convex polyhedra as Zalgaller solids.) In the notation, I will simply refer to as JZ solids. This book does not list all of these 92 solids, but the first 15 JZ solids are shown in Fig. 4.5.

To summarize the above, the convex polyhedra consist of five regular polyhedra, 13 types of semi-regular polyhedra, Archimedean n-sided prism, Archimedean n-sided antiprism, and other JZ solids from J1 to J92. The cube is an Archimedean

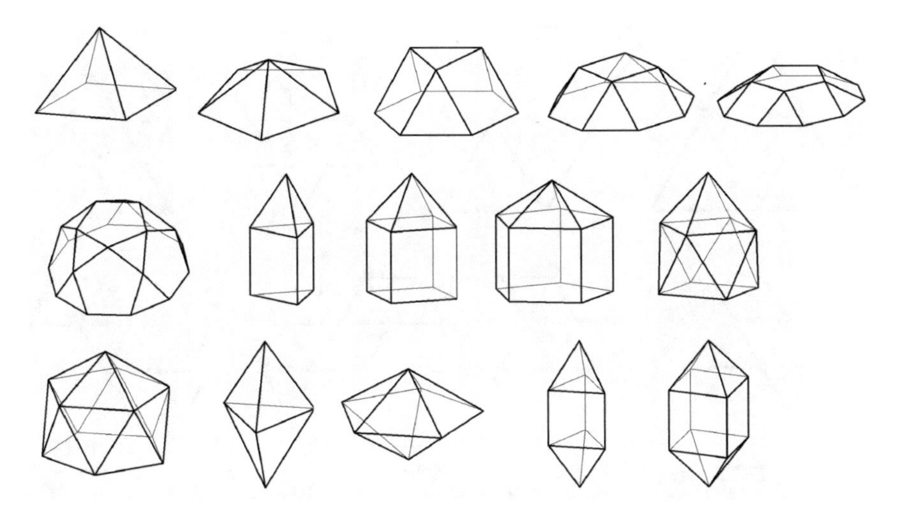

Fig. 4.5 Examples of Johnson–Zalgaller solids

4-sided prism, and the regular octahedron is also the Archimedean three-sided antiprism.

4.2 Impossibility of Common Edge-Unfolding of Regular Polyhedra

We go back to the edge-unfoldings of regular polyhedra first and then take some time to think about it. Here, we consider a problem narrowing the condition slightly to the open problem that asks if there is a common net of two regular polyhedra. That is, we consider if there is an *edge-unfolding* of a regular polyhedron that can fold into another regular polyhedron or not. Note that although it is an edge-unfolding of the original regular polyhedron, the folded regular polyhedron is not limited to its edge-unfolding. The following theorem is the conclusion of this section.

Theorem 4.2.1 *There is no edge-unfolding of a regular polyhedron that can fold into another regular polyhedron.*

Although this theorem seems to be trivial at first glance, considering the facts that a regular dodecahedron and a regular icosahedron have 43380 different edge-unfoldings, and these facts were not confirmed until 2012, it is not so trivial and unexpectedly hard to prove accurately. On the other hand, since it utilizes various (unobvious) properties of the nets, it is a good problem worth thinking to deepen our understanding of the nets.

Among the five polyhedra, the nets by edge-unfolding of the regular tetrahedron, the cube, and the regular octahedron are only $2 + 11 + 11 = 24$ in total, so it can be

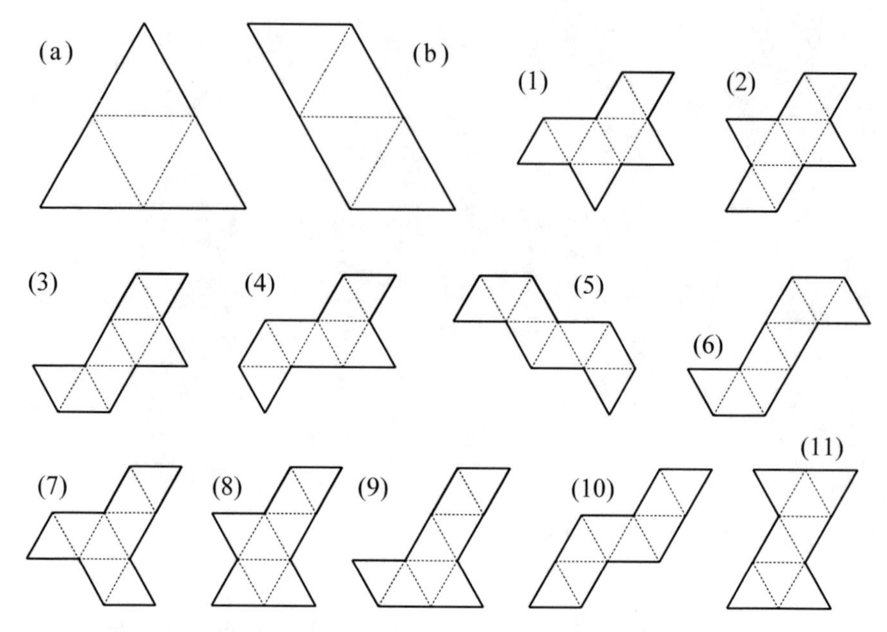

Fig. 4.6 All nets of a regular tetrahedron and a regular octahedron by edge-unfolding

predicted that they seem not to fold into other regular polyhedra with your own eyes (the 11 nets of the cube are shown in Fig. 1.1, the 2 nets of the regular tetrahedron are shown in Fig. 4.6a, b, and the 11 nets of the octahedron are shown in Fig. 4.6(1)–(11)). The suspicious ones are the regular icosahedron and the regular dodecahedron. However, the regular dodecahedron is based on regular pentagon, so it seems to be difficult to fold into another regular polyhedron. The problem would be the regular icosahedron. There are 43380 nets by edge-unfolding where 20 regular triangles are connected. Among them, for example, there may be ones that can be folded into a regular octahedron or a regular tetrahedron. Before we clear up the suspicion, we will first show a useful lemma below.

Lemma 4.2.2 *Any net of a cube, a regular octahedron, a regular dodecahedron, or a regular icosahedron is not a convex polygon. Moreover, there are at least two concave vertices.*

Here I point out that the two nets by edge-unfolding of a tetrahedron are convex polygons.

Proof For a given polyhedron, when we consider the vertices and edges of the polyhedron as the graph G as it is, the cut lines C of an edge-unfolding form a spanning tree of G by Theorem 2.1.1. Even when it is not an edge-unfolding, the cut lines C form a tree that spans whole the vertices of the polyhedron. It is known that any tree has at least two leaves (vertices with degree 1). (This is one of the basic theorems in graph theory. I recommend interested readers to read a textbook on graph theory. In a chapter of trees, there should be a proof of this fact.) Since the cut line in C ends at the vertex corresponding to this leaf, the neighborhood of this vertex is cut open as it is. Considering this surrounding angle, the cube is 270°, the regular octahedron 240°, the regular dodecahedron 324°, the regular icosahedron 300°, which are vertices that are not convex on the net. Since any tree has at least two leaves, there are at least two concave vertices. □

4.2.1 Proof of Theorem 4.2.1

Let's move on to the proof of Theorem 4.2.1. First of all, let the original regular polyhedron be Q. Then Q is either regular tetrahedron, cube, regular octahedron, regular dodecahedron, or regular icosahedron. First consider the case that Q is one of regular tetrahedra, cube, and regular octahedron. In this case, there are only 24 patterns in total (Figs. 1.1 and 4.6). Although it can be shown by formal discussions as shown later, we consider that it is almost obvious that the other regular polyhedra will not be folded from any of these 24 patterns, and hence we omit the proof for them in this book. Next, consider the case where Q is a regular dodecahedron. The regular dodecahedron is composed of 12 regular pentagonal faces. Therefore, in its net P, each angle is an integer multiple of 108°. Using this angle, it is obvious that you cannot make any angle of the vertex of 270° of the cube, the vertex of 240° of the regular octahedron, and the vertex of 300° of the regular icosahedron. Also, for a regular tetrahedron, the angle of the vertex is 180°. That is, the angles allowed for a regular tetrahedron are only 180° and 360°. However, because 108°, 216°, and 324° are the angles at a vertex of P, we cannot create such an angle, or P cannot fold into a regular tetrahedron.

Therefore, it turns out that it is enough to consider only when Q is a regular icosahedron. That is, P is a polygon consisting of 20 regular triangles, and all we have to do is show that we cannot fold a regular tetrahedron, a cube, a regular octahedron, or a regular dodecahedron from this P. Moreover, as with the above discussion, you cannot make an angle 270° of a vertex of the cube or an angle 324° of a vertex of the regular dodecahedron from P consisting only of the regular triangle. Therefore, the proof is completed if we can show that a regular tetrahedron or a regular octahedron cannot be folded from any edge-unfolding P of the regular icosahedron. We consider these two cases separately.

4.2.1.1 Proof that a Regular Octahedron Cannot be Folded from any Edge-Unfolding P of a Regular Icosahedron

To derive contradictions, we assume that the regular octahedron is folded from an edge-unfolding P of a regular icosahedron. Here, we first suppose that the length of an edge of each regular triangle of the regular icosahedron is 1, and call each triangle a unit regular triangle. Here, let ℓ be the length of the regular octahedron's edge.

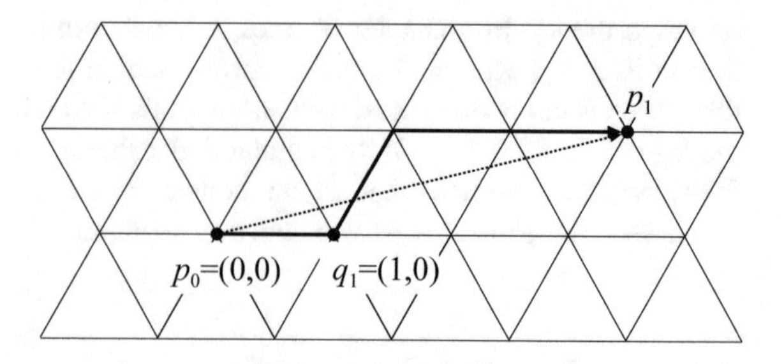

Fig. 4.7 Making the length of an edge of a regular octahedron from unit regular triangles

Then, since 20 unit regular triangles should be pasted together to make 8 regular triangles of the octahedron with an edge length of ℓ, we have $\ell = \sqrt{20/8} = \sqrt{5/2}$. Here, because P is a net of a regular octahedron and by Lemma 4.2.2, P has at least two vertices of 240°. Let p_0 and p_1 be these vertices. Since P is originally an edge-unfolding of an icosahedron, the part that realizes this 240° angle of P is formed by four corners of the unit regular triangles. Therefore, on the folded regular octahedron, the following fact holds.

Fact 1: Both vertices p_0 and p_1 of the regular octahedron folded from P are formed by four vertices of the original unit regular triangles.

We consider the vertices of the original unit regular triangles that make p_0 and p_1. We note that the vertex p_0 can correspond to at most four vertices on P, and the same thing holds for the vertex p_1. On P, we arbitrarily take any vertex p_0' corresponding to p_0, and p_1' corresponding to p_1. We consider these p_0' and p_1' on P as p_0 and p_1 again. Then we have the following fact.

Fact 2: On the net P of the regular icosahedron, we can reach between p_0 and p_1 by tracing the edges of the unit regular triangles in order.

By Fact 1, the distance between p_0 and p_1 is measured on the surface of the regular octahedron, which is ℓ or 2ℓ. By Fact 2, we can see that this distance is realized by pasting unit regular triangles and tracing their edges between them.

In other words, intuitively, p_0 and p_1 should be the two grid points on the unit regular triangular lattice, such as Fig. 4.7. Let $p_0 = q_0, q_1, \ldots, q_k = p_1$ be the sequence of vertices that appear on any shortest path formed by the edges of the unit regular triangles joining p_0 and p_1. Now, consider $p_0 = q_0$ as the origin $(0, 0)$, and we can assume that $p_1 = q_k$ is in the first quadrant of the coordinate plane and $q_1 = (1, 0)$ without loss of generality. Here, when we define each vector $\vec{q_i} = q_i - q_{i-1}$, we can regard as $p_1 = \vec{q_1} + \cdots + \vec{q_k}$. Since vectors are commutative, by exchanging them appropriately, after all, we can represent as

$$p_1 = i(1, 0) + j\left(\frac{1}{2}, \frac{\sqrt{3}}{2}\right)$$

for certain natural numbers i and j. Now we remind that the distance from the origin to this point p_1 was ℓ or 2ℓ. In the former case, we have

$$\left(i+\frac{j}{2}\right)^2+\left(\frac{\sqrt{3}j}{2}\right)^2=i^2+j^2+ij=\left(\sqrt{\frac{5}{2}}\right)^2=5/2.$$

Since i and j are natural numbers, i^2+j^2+ij cannot be equal to $5/2$. On the other hand, in the latter case

$$\left(i+\frac{j}{2}\right)^2+\left(\frac{\sqrt{3}j}{2}\right)^2=i^2+j^2+ij=\left(2\sqrt{\frac{5}{2}}\right)^2=10.$$

Here, when both i and j are odd numbers or when either one is odd, i^2+j^2+ij is odd; hence, it does not become 10. However, when both i and j are even numbers, i^2+j^2+ij is a multiple of 4; hence, it does not become 10 again. Therefore, there are no natural numbers i and j satisfying the condition. This is a contradiction.

From the above discussion, the regular octahedron cannot be folded from any edge-unfolding of the regular icosahedron.

4.2.1.2 Proof that a Regular Tetrahedron Cannot be Folded from any Edge-Unfolding *P* of a Regular Icosahedron

As with Sect. 4.2.1.1, let us assume that a regular tetrahedron can be folded from an edge-unfolding P of a regular icosahedron to lead contradictions. At this time, setting ℓ again for the length of the edge of the tetrahedron which is folded, then $\ell=\sqrt{20/4}=\sqrt{5}$. Unlike the regular octahedron, since the angle surrounding the vertex of a regular tetrahedron is 180°, we need to devise to prove. In particular, we have to pay attention that we can fold a tetramonohedron which is close to a regular tetrahedron as shown in Fig. 2.8.

First, we consider if a rolling belt introduced in Sect. 2.3.2 can appear or not. Suppose that a rolling belt appeared on the way of folding the regular tetrahedron. In this case, since the rolling belt is made of a sequence of unit regular triangles, the total length of the belt is an integer. Here, the rolling belt would make two vertices of the regular tetrahedron, and the angle of the paper around each vertex was 180°. That is to say, this alone does not contradict the regular tetrahedron. However, the rolling belt makes half the length of the full length and makes vertices at both ends. In this case, the half-length should be $\ell=\sqrt{5}$, which contradicts that the total length of the belt is integer. Therefore, we can assume that the rolling belt does not appear on the way to fold the regular tetrahedron. In other words, if the regular tetrahedron can be folded from an edge-unfolding P of a regular icosahedron, it can be assumed that a rolling belt does not appear halfway.

Now we consider P as a net of the regular icosahedron. By Lemma 4.2.2, P has at least two vertices of 300°. Let one of them be p_0. The point p_0 cannot ultimately be the vertex of a regular tetrahedron, so this part needs to be 360°. That is, one corner of another unit regular triangle is bonded to this point p_0. This will be easy to be convinced by looking at Fig. 2.8. We denote this corner by p_0'. Then, between p_0 and p_0' should be filled with a sequence of edges of unit regular triangles. Let k denote the length of this sequence. Since P is a polygon spread out on the plane, at least one vertex of the regular tetrahedron must be folded on this sequence. This vertex must be an endpoint or a midpoint of an edge of a unit length in the sequence of edges joining p_0 and p_0'. Let c_0 be this vertex. On the other hand, a rolling belt did not appear on the way to fold the regular tetrahedron. Therefore, when we fold a regular tetrahedron from P, it is necessary to begin to glue from this vertex c_0 and glue edges of length $1/2$, which are obtained by subdividing unit edges, in order.

Let p_1 be the other vertex of 300° of P. By the same argument, we can conclude that the other vertex c_1 of degree 180° of the regular tetrahedron also corresponds to an endpoint or a midpoint of a unit edge. This is almost the same as Hirata half-lengths theorem indicated in the literature [DO07, Sect. 25.3.3].

From the above discussion, if the tetrahedron can be folded from P, the same argument as the regular octahedron can be applied with respect to the two vertices c_0 and c_1 with half of the unit length of the unit regular triangle. That is,

$$\left(\frac{i}{2}+\frac{j}{4}\right)^2+\left(\frac{\sqrt{3}j}{4}\right)^2=(\sqrt{5})^2=5$$

should be established for certain natural numbers i and j. To simplify the equation, we have $i^2+j^2+ij=20$. From the same argument as the regular octahedron, both i and j need to be even natural numbers, but there are no even natural numbers that satisfy this expression, which is a contradiction.

Therefore, the regular tetrahedron cannot be folded by any edge-unfolding of the regular icosahedron.

This is the end of the proof of Theorem 4.2.1. Even if it seems easy at first glance, it turns out that the proof on the net is not that easy. Especially since the angle around the vertex of the regular tetrahedron is 180°, it is surprisingly hard to show that it cannot be folded.

4.3 Common Nets of Regular Tetrahedron and Cube

In this section, we introduce the recent result which seems to be the most proximate to the common net of two regular polyhedra. Specifically, we introduce an algorithm that actually constructs common nets of a cube and an (almost) regular tetrahedron. Hereafter, in this section, when we say a cube, it always refers to a unit cube of size $1 \times 1 \times 1$. Let ℓ denote the length of the regular tetrahedron with the same surface area of the cube. Since the unit cube has a surface area of 6, we have $\ell = \sqrt{2\sqrt{3}}$. First, I give a very rough idea of the algorithm. We always deal with a net of a cube, which is also a net of a tetramonohedron. By Theorem 2.3.2, it means that we always consider a p2 tiling. Then, the algorithm deforms this common net gradually and brings the tetramonohedron closer to the regular tetrahedron. In this algorithm, the lengths of the tetramonohedron's edge can be adjusted by two independent parameters. This algorithm gives a positive answer in a sense to the above-mentioned open problem. Given an extremely small error $\epsilon > 0$, our algorithm halts and generates a common net of a cube and an almost regular tetrahedron such that all edges of the tetrahedron are in the interval $[\ell - \epsilon, \ell + \epsilon]$.

In the unfolding generated by this algorithm, we note that connectivity is not guaranteed in general. In other words, if you choose two parameters arbitrary, you may have disconnected unfolding, which is not a common net anymore. From the experimental results, we know how to select parameters so that the algorithm outputs connected nets. Based on the computer experiment, we chose proper parameters, and we have a common (connected) net of an "almost" regular tetrahedron and a cube whose error is $\epsilon < 2.89200 \times 10^{-1796}$ so far. The mathematical proof that this algorithm works for arbitrary small $\epsilon > 0$ is not yet solved. In addition, if we try to generate a common net of a regular tetrahedron and a cube of error 0 in this way, an infinite number of points are generated on the net. In our definition, a net should be a connected simple polygon with finite number of vertices. Therefore, we cannot call such "a set of infinite number of points" as "(polygonal) net". The arguments about figures defined by these infinite points are required separately. In this book, we do not go into this issue anymore.

There is a close relationship between the value of the parameter that adjusts the length of the edge and the pattern that appears in the boundary of the net (though we have no proof). Specifically, it is obvious between representation by the continued fraction expansion of the parameter value and repetitive patterns appearing in the boundary of the net. Although no mathematical proof for this relationship has been given at the moment, we can also use this relationship for real numbers that can be represented by infinite repetitive continued fraction expansion of simple values. Concretely, if we apply the procedure shown in this section to real numbers with such a simple continued fraction expression, we can easily generate so-called *fractal curves*.

4.3.1 Procedure for Generating Common Nets

Let's start with the net P_1 of a cube shown in Fig. 4.8. It can be understood by looking carefully that the polygon shown by the bold lines in the figure is the net of the cube. Here, the points $c_1, c_2, c_3, c_4, p, p'$ are the midpoints of the corresponding edges, and three line segments c_1c_2, L_1, and L_2 are parallel. We first confirm that P_1 satisfies the conditions of Theorem 2.3.2. If you prepare copies of this net and let

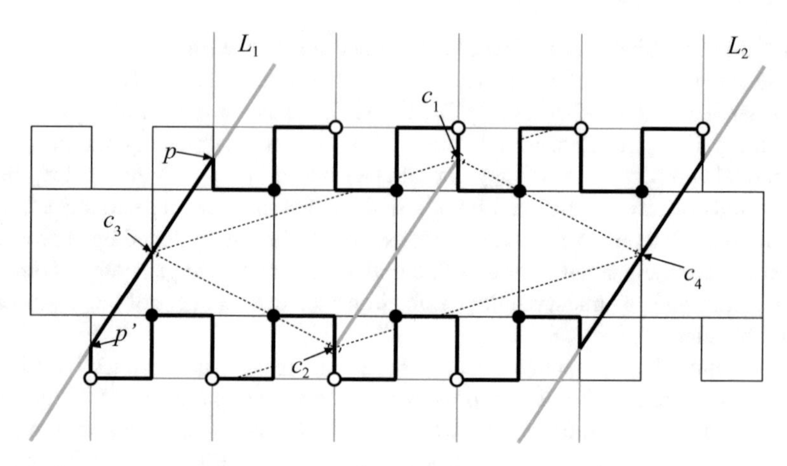

Fig. 4.8 The first net P_1 of a cube

two points c_1, c_2 as the rotation centers, you can tile copies one after the other up and down. Since c_1c_2, L_1, and L_2 are parallel, an elongated upward sloping band is created. Accordingly, you can paste copies in two other c_3 and c_4 as well, and you can fill the plane. That is, this polygon is not only the net of a cube, but also a net of a tetramonohedron. Considering a bit further, you can see that the polygon always forms p2 tiling as long as c_3 and c_4 are in symmetrical positions. In other words, this forms a rolling belt introduced in Sect. 2.3.2, and the rotation centers c_3 and c_4 of p2 tiling can move to any positions on L_1 and L_2 while maintaining symmetry. Namely, P_1 is a common net of a cube and infinite kinds of tetramonohedra. This fact may be hard to understand only with a mathematical explanation; however, you will see how easy it is by actual work.

Now, $|c_1c_2| = \sqrt{13}/2 = 1.80278$, and the area of the four congruent triangles is $3/2$. Therefore, taking c_3 and c_4 on L_1 and L_2 in Fig. 4.8 that $|c_3c_1| = |c_3c_2| = |c_4c_1| = |c_4c_2|$, we obtain the following lemma.

Lemma 4.3.1 *There exists a common net of a cube and a tetramonohedron that consists of four copies of a isosceles triangle of edge lengths* $\sqrt{13}/2 : \sqrt{745/208} : \sqrt{745/208} = 1.80278 : 1.89255 : 1.89255$.

The tetramonohedron obtained in Lemma 4.3.1 is close to a regular tetrahedron. The goal is to make the lengths of these edges closer to the same value $\sqrt{2\sqrt{3}} = 1.86121$.

We here give the procedure to transform this P_1. More precisely, this procedure is to extend the distance between two points c_1 and c_2 by moving c_1 of Fig. 4.8 to the right and c_2 to the left a little. During this deformation, two properties will not be changed. We assert these two properties as two invariants:

1. P_1 is a net of the unit cube after deformation.
2. P_1 is a net of a tetramonohedron and $|c_1c_3| = |c_1c_4| = |c_2c_3| = |c_2c_4|$ always holds for the lengths of the edges.

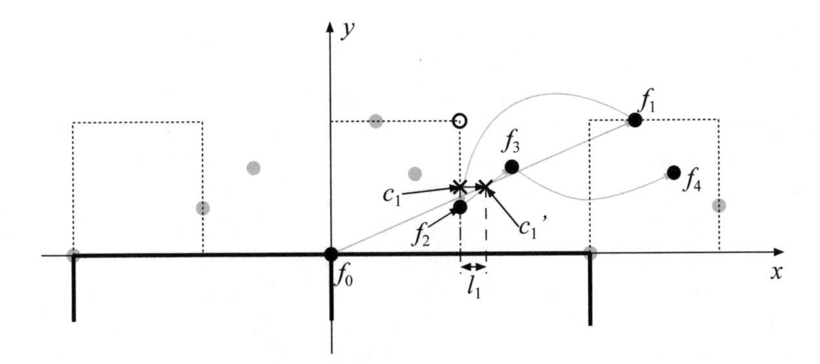

Fig. 4.9 How to construct points on the boundary of the common net of a cube and a tetramonohedron

Assuming that deformation is carried out while maintaining these two invariants, if $|c_1c_2|$ becomes $\sqrt{2\sqrt{3}}$, a common net of a cube and a regular tetrahedron is obtained.

Now, in the net P_1 of Fig. 4.8, we will move the points while maintaining the invariants, but there are "some points that must be on the net" during deformation. On the other hand, if we move the points c_1 and c_2, we have to move the positions of some points from the constraint that it should be p2 tiling. Then it causes a chain reaction, resulting in a process that generates several discrete points. As we will show later, if we change the distance $|c_1c_2| = \sqrt{13}/2$ between two points directly to $|c_1c_2| = \sqrt{2\sqrt{3}}$, this point generation process does not halt. On the other hand, if we try to stop the point generation process at a finite number of times, it causes errors. Therefore, we here have to choose to make either a common net as the limit of an infinite number of point sets or a common net with a small error $\epsilon > 0$ with a finite number of line segments. In this section, for a given error $\epsilon > 0$, we define an *almost regular tetrahedron* by a tetramonohedron whose edge length $|c_1c_2|$ is in $[\sqrt{2\sqrt{3}} - \epsilon, \sqrt{2\sqrt{3}} + \epsilon]$ and proceed to our discussion. Hereafter, we concretely show how to stretch the distance $|c_1c_2|$ between two points c_1 and c_2 from $\sqrt{13}/2 = 1.80278$ to $\sqrt{2\sqrt{3}} = 1.86121$. Intuitively, we move c_1 and c_2 away "little" in the horizontal direction in Fig. 4.8. We consider the ripple effect that occurs as a result of this stretch.

Here, looking at the ∘ of Fig. 4.8, we can see that these points come to the center of the lid or bottom of the cube. If these points are removed from the net, the face of the cube gets a hole. On the other hand, if this point comes inside of the net, it will overlap on the surface of the cube. Therefore, considering invariant (1), these points have to be on the boundary of P_1. On the other hand, the • of Fig. 4.8 is the vertex of the cube, and it must also be on the boundary of P_1 again for the same reason. Let's call these points that we cannot move from the boundary of the net as *fixed point* of the net.

We consider only the upper half of P_1 for a while. There are eight fixed points and rotation center c_1 in the upper half of P_1. We tentatively remove all other points.

In other words, at this point, the points on the boundary of the net are only the fixed points and the rotation center. We will add the points on the boundary of the net in order, and finally generate a net that is obtained by connecting these points with line segments. For accuracy, put the xy-coordinate axis on the edge. Let f_0 be the fixed point corresponding to the vertex of the cube at the lower left of c_1 as shown in Fig. 4.9, and let the coordinate of f_0 be the origin $(0, 0)$. Since the length of the edge of the cube is the unit length, the coordinate of c_1 is $(1/2, 1/4)$. Here, we take a certain small value ℓ_1 as the shift width, and assume that the rotation center $c_1 = (1/2, 1/4)$ is moved to $c_1' = (1/2 + \ell_1, 1/4)$. Then since f_0 is the fixed point and c_1' is the new rotation center, the point $f_1 = (1 + 2\ell_1, 1/2)$ on the opposite side of f_0 with respect to c_1' is required to be a point on the boundary of the net. Here, remembering that this is a net of a cube, the point f_1 is glued to the point $f_2 = (1/2, 2\ell - 1)$ when it is folded into a cube. Therefore, this f_2 must be a point on the boundary of the net. Besides since this net of the cube opens symmetrically, we can define an equivalence relation $(x, y) \equiv (x - 1, y)$ for any (x, y) and $(x - 1, y)$. Therefore, if the point $f_2 = (1/2, 2\ell - 1)$ is on the boundary of the net, its equivalent points such as $(1 + 1/2, 2\ell_1)$ are also on the boundary. (This equivalence relation shall be applied appropriately hereinafter.) In this way, when this mapping is applied iteratively from the eight fixed points and the new rotation center $c_1' = (1/2 + \ell_1, 1/4)$, the set of the points which must be on the boundary of the net can be computed. In this procedure, the following lemma holds.

Lemma 4.3.2 *The above fixed point projection procedure halts at a finite number of times if and only if the shift width ℓ_1 is a rational number.*

Proof When ℓ_1 is a rational number $\frac{p}{q}$ for some relatively prime natural numbers p and q with $0 < p < q$, the set of possible points generated by the projection procedure is contained in points on the grid of size $O(pq)$. Therefore, if we repeat the projection procedure $O(pq)$ times, the procedure eventually visits the same point twice, and the point set will be fixed there. On the other hand, if ℓ_1 is not a rational number, the same coordinates will never appear again, so the procedure will not terminate. □

With the above procedure, we can shift c_1 and c_2 independently in the horizontal direction and extend $|c_1c_2|$. A specific method of constructing the net is as follows. First, set two appropriate shift widths for c_1 and c_2 and shift them in the net of Fig. 4.8. Next, compute all the points on the boundary of the net with the above procedure (independently for c_1 and c_2). Then, connect the points with straight lines to establish the upper boundary and the lower boundary of the net. Finally, redraw L_1 and L_2 so that they become parallel to the shifted result c_1c_2 and the surface area does not change. For example, Fig. 4.10 is a common net of a cube and an almost regular tetrahedron obtained by setting $\ell_1 = 4/21$ and $\ell_2 = 5/24$.

This procedure seems to be all going well, but in fact, we cannot guarantee the connectivity of the net in this general way. Specifically, when we draw two line segments L_1 and L_2, the net may be disconnected. A more qualitative analysis is necessary for the boundary of the net generated with the above procedure in order to ensure

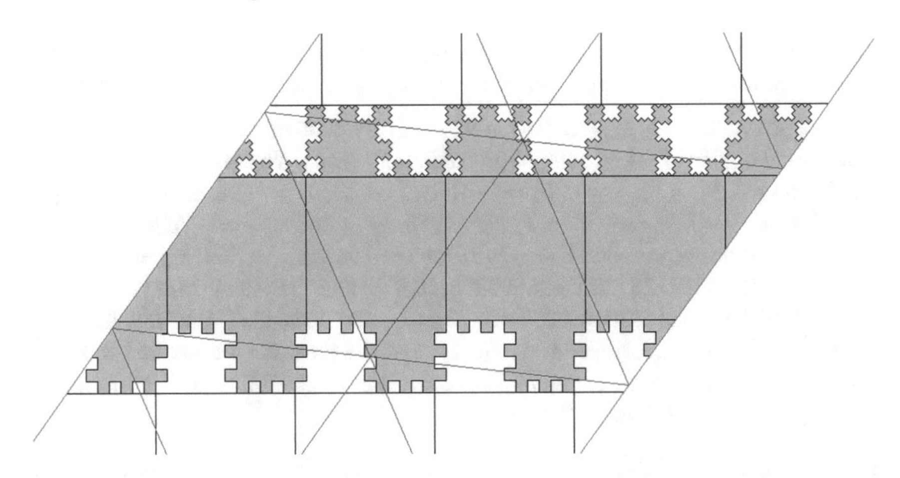

Fig. 4.10 An example of a common net of a cube and an almost regular tetrahedron

that the net is not divided even if these two line segments are drawn. The following observation is obtained by numerous experiments, but formal characterization and mathematical proof are still open.

Observation 4.3.3 *Let* $\phi_1 = \frac{1}{a_1\pm}\frac{1}{a_2\pm}\frac{1}{a_3\pm}\cdots\frac{1}{a_k}$ *and* $\phi_2 = \frac{1}{b_1\pm}\frac{1}{b_2\pm}\frac{1}{b_3\pm}\cdots\frac{1}{b_h}$ *be the result of continued fraction expansions of the two rational numbers* ϕ_1 *and* ϕ_2.[1] *For these two rational numbers, let* $\ell_1 = (1-\phi_1)/4$ *and* $\ell_2 = (1-\phi_2)/4$. *Then, the upper half of the net is obtained by recursively replacing each line segment in zigzag "wavy line" in order by the value of* a_i *(see Fig. 4.10). Precisely, this wavy line is a "square wave" if* a_i *is an even number and "triangle wave" if it is an odd number. The number of zigzag peaks is given by the value of* a_i. *The direction in which each wavy line goes first is determined by the sign of* a_i. *The same applies to the lower half.*

For example, in Fig. 4.10, $\ell_2 = (1-1/6)/4 = 5/24$, that is, $\phi_2 = 1/6$. When we plot the points using this value and construct the bottom half of the net, the edge of P_1 is replaced by a square wave with six peaks, matching with $1/6$ which is the continued fraction expansion of ϕ_2. At the same time, $\phi_1 = 5/21 = 1/(4+1/5)$ in the upper half. In the net constructed by plotting points, the zigzag wave is obtained by replacing each edge of P_1 with a square wave with four peaks first, and then each line segment is replaced with a triangular wave with five fine peaks.

As mentioned earlier, Observation 4.3.3 is only a conjecture or an explanation of experimental phenomenon, and the feature and proof of this "wavy line" are unresolved. However, assuming that Observation 4.3.3 holds, it is possible to construct

[1]Precisely, the "continued fraction expansion" used here is slightly different from the standard continued fraction representation. In the standard continued fraction expansion, $a_i \geq 1$ for each term a_i, all signs are "+". On the other hand, our continued fraction expansion here also uses the sign of "−", so that $a_i > 1$ holds when $i > 1$.

a connected net. Specifically, exploring values that do not disconnect the net, we determined the values in the order of $a_1, b_1, a_2, b_2, \ldots$ and made the length closer to the target value $\ell_1 + \ell_2 = \sqrt{2\sqrt{3} - 9/4} - 1$. By constructing a full search algorithm based on this idea and computing the first 50 terms, we obtained the following values: $a_1 = 4, b_1 = 6, a_2 = 6, b_2 = -34, a_3 = -42, b_3 = -14, a_4 = -116, b_4 = -2146, a_5 = 4010, b_5 = -3316, a_6 = -4958, b_6 = 8684, a_7 = -7820, b_7 = 7082, a_8 = 2668, b_8 = -3684, a_9 = 4564, b_9 = 1662, a_{10} = 560, b_{10} = -158, \ldots, a_{49} = -90102250121781322369882246$, $b_{49} = 43038114499907513142235 6706$, $a_{50} = -1990926111818830489021958 58$, $b_{50} = -4002838642627588575988519794$. If you adopt $a_1, a_2, \ldots, a_{10}$ and $b_1, b_2, \ldots, b_{10}$, the error is $\epsilon < 4.63451 \times 10^{-56}$, and if you adopt all up to a_{50} and b_{50}, the error is $\epsilon < 2.89200 \times 10^{-1796}$. From the above, we have the following theorem.

Theorem 4.3.4 *For the error* $0 < \epsilon < 2.89200 \times 10^{-1796}$, *there is a common net of a cube and an almost regular tetrahedron of error* ϵ.

Of course, there is no special meaning for 50, and if we apply this algorithm to larger numbers, we may reduce the error as much as possible.

Conjecture 4.3.5 *For arbitrary* $0 < \epsilon$, *there is a common net of a cube and an almost regular tetrahedron of error* ϵ.

4.3.2 Summary of This Section and Future Work

In this section, we showed how to generate a common net of a cube and an almost regular tetrahedron. However, in our procedure, an infinite number of points is necessary to generate a common net of a cube and an accurate regular tetrahedron. Whether these points converge or not is unknown, but even if they converge, there is room for discussion as to whether this figure can be called a "net". On the contrary, there may be arguments that there is no common net of a cube and a regular tetrahedron because there are infinitely many points required. In any case, more detailed discussion is needed for the point set generated by the procedure proposed in this section and the net defined by this set of points. We may be able to construct a common net of other regular polyhedron and regular tetrahedron in the same way.

Finding a common net among regular polyhedrons other than regular tetrahedron seems to be more difficult problem that remains open because characterization by the tiling (Theorem 2.3.2) cannot be used anymore. Intuitively, it seems unlikely that such common nets exist. However, as there is a beautiful net in Fig. 4.1, we consider that it is not so easy to show negative results.

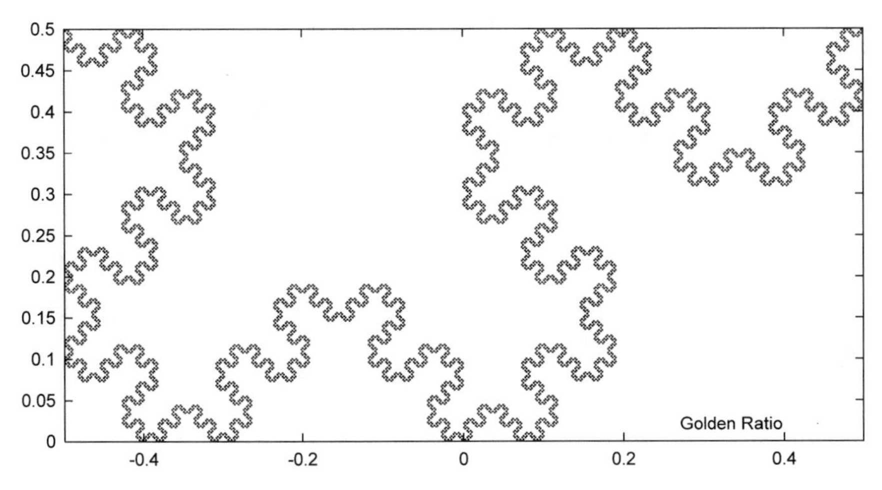

Fig. 4.11 Plotting 5000 points based on the golden ratio $\phi = \frac{1+\sqrt{5}}{2} = 1 + \frac{1}{1+}\frac{1}{1+}\frac{1}{1+}\cdots$

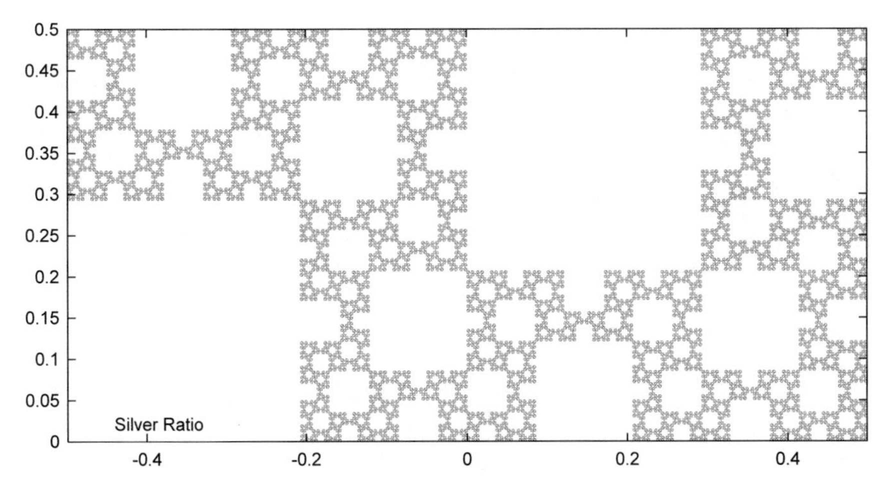

Fig. 4.12 Plotting 10000 points based on the silver ratio $\phi = \sqrt{2} - 1 = \frac{1}{2+}\frac{1}{2+}\frac{1}{2+}\cdots$

4.3.3 Bonus

Irrelevant to the net, assuming Observation 4.3.3, if you select the real number ϕ suitably, you can easily generate fractal patterns. Two execution examples are shown in Figs. 4.11 and 4.12.

Reference

[DO07] E.D. Demaine, J. O'Rourke, *Geometric Folding Algorithms: Linkages, Origami, Polyhedra* (Cambridge University Press, Cambridge, 2007)

Part III
Algorithm and Complexity of Folding

In Part III, we turn to a simple origami model away from the solid. When it comes to origami, it is a molding that folds two-dimensional squares with various folding lines. In recent complex origami, extremely fine and accurate folding such as pleat folding in 128 equal parts is required. When making such equally spaced creases, you notice that the number of folds varies depending on the way of folding. Specifically, if you fold paper in two or more sheets once, you can save the number of folds. On the other hand, if you fold it in a lot of sheets at the same time, it will be disturbed by the thickness of the paper, and you will not be able to fold with accuracy.

Useful frameworks for thinking about such "how to fold" are " *algorithm* " and " *computational complexity* ". They are the frameworks for discussing the skill of "method for computation" that comes out in computer science. The idea of applying these frameworks to origami is a recent trend in the field of computational geometry.

First, algorithm is the method of computation. Although it seems inseparable with computer programs, in fact, the history of the algorithm itself is much older than computers. "Euclidean algorithm" is a procedural method for finding the greatest common divisor of two natural numbers.[1] This is said to be the oldest algorithm which is known from about 300 BC. Since the real computer was invented in the 1930s, it can be said that the idea of algorithm itself is over than 2000 years older than computers.

On the other hand, computational complexity is the cost necessary for computation. It is used when quantifying the cost needed to compute and discussing goodness or badness of algorithms. In the case of computers, there are *time complexity* to evaluate computation time and *space complexity* to evaluate the amount of memory required for computation. Generally, it is known that time and space are in a trade-off relationship with algorithms on computers. In other words, generally, fast algorithms tend to consume a lot of memory and save computation speed when saving memory. Of course, developing a clever algorithm sometimes makes it possible to create a high-speed program with reduced memory. When introducing this approach into

[1]Euclidean algorithm is a computation procedure (.=.algorithm) for finding the greatest common divisor for two given natural numbers p and q (for the sake of simplicity, we assume $p > q$). Let r be the remainder obtained by dividing p by q. If r is 0, then q is the greatest common divisor. If r is not 0, repeat the same operation regarding as q and r are new p and q. If we repeat this until r becomes 0, then the last q is the greatest common divisor of the original p and q.

origami, we can discuss the computational complexity of origami. For example, even with the same origami model, if they had to fold 500 times or 50 times, most people would like the latter. On the other hand, folding a lot of paper will be difficult if the number of sheets increases, and precision will also deteriorate, so you would want to avoid this situation. In order to discuss such an intuitive feeling more quantitatively and theoretically, research on origami algorithm and computational complexity has started.

In Part III, we will learn the latest results of "computational complexity" and "algorithm" of computational origami, which is a new academic field.

In any framework, in order to consider algorithms and computational complexity, it is necessary to properly define the "model" as the basis of the discussion and the "basic operations" allowed on the model. In other words, to discuss the quality of algorithms, it is meaningless unless considering the same set of basic operations on a common model. Such a model and the basic operations on the model must be valid and persuasive in the origami field. In the origami society, basic operations called "Huzita's six axioms" and "Hatori's seventh axiom" have already well known, and this framework is quite stable. (Since these basic operations are not handled directly in this book, concrete axioms are omitted here. Interested readers should refer to the reference [DO07].) Therefore, in considering general origami algorithms and computational complexity, it would be reasonable to premise these basic operations on a two-dimensional plane. Given such a background, "origami" and "algorithm" are compatible.

In computer science, the efficiency of an algorithm is measured by the time complexity as the number of basic operations to be performed and the space complexity as the memory area to be used. What is equivalent to such "time complexity" or "space complexity" in "origami algorithm"? Regarding the time complexity, "number of folds" based on the basic axioms is considered as a natural correspondence. This has the name *folding complexity*. On the other hand, what about space complexity? In recent years, the notion of crease width has been proposed, and research has begun. Nonetheless, these notions are still not absolute and research has just begun.

Here, we introduce models of "origami algorithm" and "computational complexity of origami" and the latest results. First, I clarify "model": "origami" handled in this section is a one-dimensional line segment. In other words, for example, imagine that the origami is an elongated paper tape with vertical creases. Moreover, most of the current results are the results of model with equally spaced creases. Yes, it is a very simple model that cannot be further simplified. From the viewpoint of algorithm and computational complexity, although it is such a simple model, there are many problems that have not been solved yet, and it is a surprisingly deep theme. Of course, the following extensions will be easy to consider:

- Extension to non-equidistant creases.
- Extension to 2D plane.
- Extension to diagonal creases.

At the time of writing this book, it is the actual situation that we cannot reach such extensions. Conversely, it is an unexplored area where there is still room for cultivation. We will introduce these undeveloped areas in the latter part of Part III and Part IV. In the field of computational origami, it is not uncommon that discovery of college students, and in some cases, high school students, tends research to make progress. While experiencing the interest of computational origami as applications of computer science, you might want to think about what to do about various studies.

Computational complexity in origami

Both of the notions "folding complexity" and "crease width" introduced in this book are invented by me and named by my collaborator Erik D. Demaine, who is a professor at MIT, and one of his favorite research topics is computational origami. (In fact, he is the godfather of the research area "computational origami"!) At a time when both notions were presented and discussed, the results gradually came out, and they were aptly named. Since they have good compatibility with the philosophy of computer science in various viewpoints, research of these notions is expected to make progress in the future. In the past, several attempts at research of "cost of folding" had been made. For example, there was some research to estimate the cost of origami design by the summation of the cost which corresponds to each folding. However, it seems that none of them has been established. As far as I know, it is the first time that we investigate the complexity of origami from the viewpoint of computer science in this book.

Chapter 5
One-Dimensional Origami Model and Stamp Folding

Abstract In this chapter, we first clarify the model and problems of one-dimensional origami that we will handle. Roughly, our origami is a long rectangular strip, and crease lines are orthogonal to the long side of the strip. That is, they are parallel to each other. Moreover, these crease lines are placed at regular intervals on the strip. As you can imagine, this is the simplest origami model in one-dimensional. In such a simple model, we have many problems from the viewpoint of algorithms.

5.1 One-Dimensional Origami Model with Equivalent Spaced Creases

In this chapter, we mainly deal with the following problem.

Input: A paper strip of length $n + 1$ is equally spaced with n assignments of M and V. The assignments M and V represent a mountain fold and a valley fold, respectively.

Output: The paper strip folded into length 1 according to the given assignments.

We ignore the paper thickness in this section. Considering the implementation on a computer, it is easy to handle by thinking in the following way.

We can regard the input as a string of length n consisting of M and V. A bit more formally, the input of the problem is the string s of length n on the alphabet $\{M, V\}$. The output needs to be considered a little more. We first consider that a paper strip of length $n + 1$ is formed by the sequence of $n + 1$ unit length segments. Then, the portion from the edge of the paper to the first M or V is segment 0, the next is segment 1, and the last one is segment n. At this time, the folded state of the paper strip can be represented by the arrangement of the overlapping segments in order from the top. We have to take care of the upside or downside of each segment. Here we assume that the segment 0 does not change the initial position and direction without loss of generality. For example, consider that $s = VVV$ is given as a crease pattern on a paper strip of length 4. Then, there are three different folded states as shown in Fig. 5.1. We denote them by [1|3|2|0], [1|0|3|2], and [3|1|0|2], respectively. In this

R. Uehara, *Introduction to Computational Origami*,
https://doi.org/10.1007/978-981-15-4470-5_5

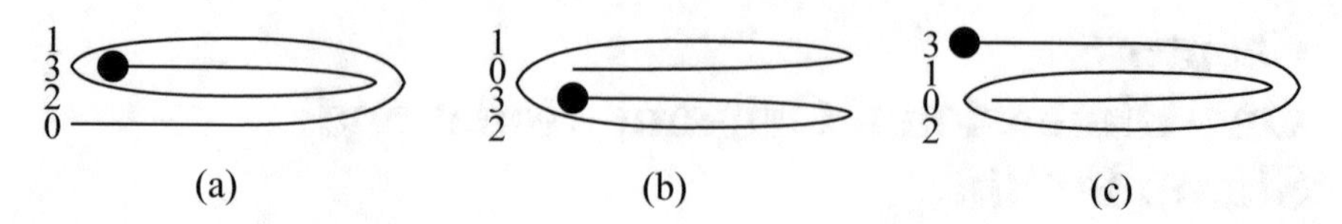

Fig. 5.1 The folded states for an M/V assignment VVV. Each • represents the end of the segment n

way, for a given crease pattern, there are many folded states which realize the same crease pattern in general. We consider some problems here.

- Does folded state exist for any M/V assignment?
- What kind of M/V assignments do that have less or many folded states?
- Generally, are there many folded states for a given M/V assignment?
- If there are a lot of folded states, what kind of folded states are "good" or "bad"?

Before tackling these problems, we prepare a couple of things.

5.1.1 Basic Theorems

There are two properties, which can be understood immediately. Both are simple but important, so we establish them as theorems.

Theorem 5.1.1 *For any M/V assignment, there exists a folded state that realizes the assignment.*

Proof Consider the end of the paper strip, that is, the segment n. One edge of the segment is the edge of the paper strip, and the other edge is assigned by M or V. Then, fold the segment n according to this assignment, glue it, and remove it from the paper strip on the operations after that. This is equivalent to reduce a paper strip of length $n-1$. Therefore, if we repeat the same operation, we can eventually fold it down to length 1 (since thickness is now being ignored). Therefore, there is a folded state that realizes the given assignment. □

Theorem 5.1.2 *The following three are equivalent: (1) The input string s is an alternating string of M and V, namely, $MVMVMV\cdots$ or $VMVMVM\cdots$. (2) The folded state is the pleat folding (Fig. 5.2). (3) There is only one folded state for the M/V string s.*

Proof Here, it is obvious that (1) and (2) are equivalent. Rather, (1) can be said to be the definition of the pleat folding of (2). We show that they are equivalent to (3). First, we show that (1) implies (3). If you fold it repeatedly from the edge, the paper strip is folded alternately, and there is only one way to fold. Next, we show that (3) implies (1) by contradiction. If s does not satisfy the condition of (1), without loss

Fig. 5.2 An example of pleat folding

of generality, s contains two consecutive mountain folds MM. When we fold both sides of this MM in the same way as the proof of Theorem 5.1.1, we obtain two mountain folds on the paper strip of length 3. Since this paper strip of length 3 has two folding states, it contradicts the condition of (3). Therefore, when two creases of the same direction are consecutive, there are two or more folded states. Therefore, (3) implies (1), which completes the proof. □

By Theorem 5.1.2, patterns other than pleat folding have at least two folded states. Then does the pattern "close" to pleat folding have less folded states? This is surprisingly difficult. Consider the following problem as an example.

Exercise 5.1.1 Let s' be a pleat pattern $MVMV \cdots M$ of length n for odd number n. Now we connect two s's and obtain $s = s's' = MVMV \cdots MV\underline{MM}VMV \cdots M$ of length $2n$. This string s is an almost pleat pattern such that the pattern MM appears only in the central part. Then how many folded states of this pattern are there?

At first glance, it seems to be less, but in fact, it can show that there are exponentially many folded states. Let me give you an example that will reappear later on as a pattern with interesting properties.

Example 5.1.3 Let $s = MMVMMVMVVVV$ be the M/V assignment for a paper string of length 12. Then there are 100 ways to fold this M/V assignment to unit length. That is, this pattern has 100 folded states of unit length.

The fact that there are one hundred ways of folding for such a short string is surprising to many readers, isn't it?

However, you will wonder how I figured out this number 100. Actually, at the present time, there is no formula to obtain instantly the number of ways of folding that realize a given string. This example is the result of the exhaustive search by a computer program. Therefore, at the moment, "trying and counting all possible ways" is the only barely irresponsible proof. Moreover, I wrote the program to count the number of the folded states for the pattern in Example 5.1.3, but this program is not that simple algorithm. I show concrete implementation method briefly in Example 6.2.2, so think about the program if you like programming.

5.2 Universality of Simple Folding Model with Equivalent Spaced Creases

We first consider the folding model. We already mentioned that there are basic operations called "Huzita's axioms" and "Hatori's operation" as the basic operations of two-dimensional origami, but here we consider a more basic model of operations in one-dimensional origami. Specifically, we introduce a folding model called *simple folding model*. It is a model of folding operation introduced by Arkin et al. [ABD+04]. Briefly, the basic operation which allowed in the simple folding model consists of the following steps:

- First, spread the paper flat (this state is called *initial state*).
- Choose a crease line, pick up some inner paper layer(s) from the inside, and fold into a valley along the crease line (or open the folded inner paper).
- Flip the folded paper to the opposite side, make the paper flat, and repeat the folding operation.

Example 5.2.1 Consider the folded state on the left in Fig. 5.3. When we choose the crease (a), we can perform a simple folding with 1, 2, and 5 layer(s), but we cannot perform it with three and four layers. When we choose two layers, we flip them from left side to right side and obtain the folded state on the top right in Fig. 5.3. On the other hand, when we choose the crease (b), we can perform a simple folding with one, two, four, and five layer(s), but we cannot perform it with three layers. When we choose two layers, we flip them and obtain the folded state on the bottom right in Fig. 5.3. We note that this simple folding operation is in fact an unfolding operation for these two layers.

Considering practical applications such as designing automatically folding robots, it is interesting to see how far it can be achieved with such a simple folding. First, we show simple but very important theorems in this model.

Theorem 5.2.2 *You can mountain fold with simple folding model.*

Proof Turn over the paper and fold all paper layers into valley folds. □

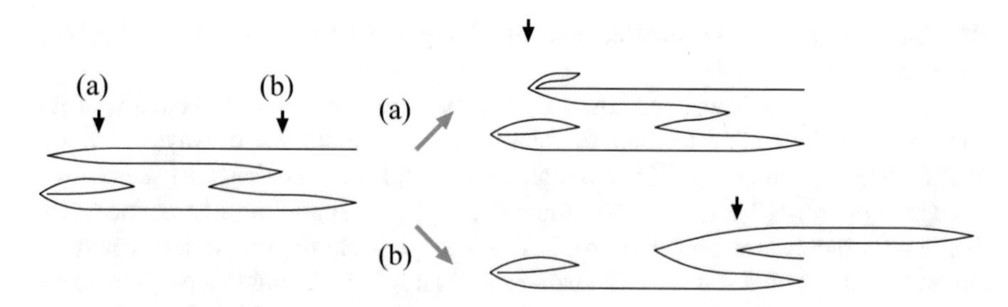

Fig. 5.3 Two examples of simple folding operations

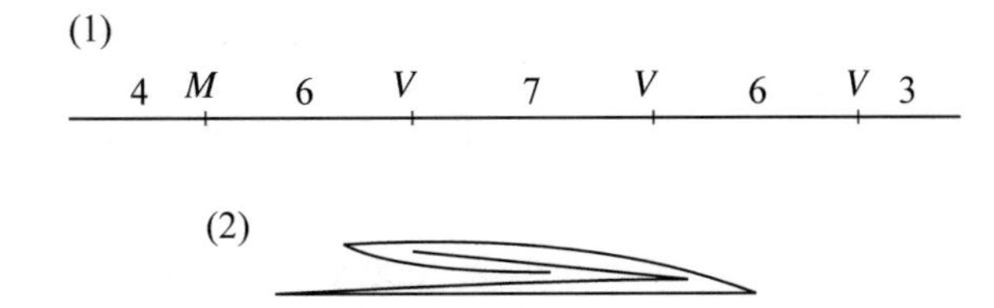

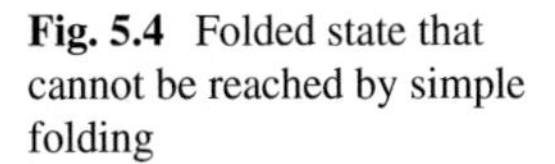
Fig. 5.4 Folded state that cannot be reached by simple folding

Theorem 5.2.3 *If a certain folded state Q can be folded by simple folding from a certain folded state P, the folded state P can also be folded by simple folding from the folded state Q.*

Proof In simple folding model, the reverse operation of a folding operation can also be regarded as a simple folding operation, so each operation can be reversed. □

Given that mountain folding can be done, it seems like an ordinary operation at first glance, and it may not be clear the difference from ordinary folding. Moreover, either theorem seems like trivial thing, but is it the case?

For example, look at the pattern shown in Fig. 5.4. From the upper flat state (1), according to the given lengths and creases, can we fold into the folded state (2) with only simple folding operations? It is impossible. Proof is easy if you notice something. If state (2) can be reached from state (1), it should be able to return from state (2) to the previous folded state by Theorem 5.2.3. However, in state (2), the paper layers are locked to each other and cannot be opened at all by simple folding. Therefore, from the state (2), it cannot transform to any other folded state, especially, it cannot reach the folded state (1). Thus, we cannot fold to the state (2) from the state (1) by simple folding.

Conversely, if any folding is allowed, arbitrary valid folded state can be realized. Although this may seem to be trivial at first glance, it was an extremely challenging problem in the case that the paper was hard and did not bend at a part other than the crease. In fact, a workshop only for this problem was held in the society of computational geometry, and it was so difficult that it took years to be settled.

Theorem 5.2.4 *Assume that the paper is hard and not folded except at the crease lines. If you allow any folding operation, you can transition from any folded state to another folded state.*

Proof Considering that paper segment is hard and not bent except creases, this is equivalent to a one-dimensional *linkage*. A *linkage* is intuitively a sequence of bars such that each length of a bar does not change, and they are connected by freely rotating joints. The problem whether linkage locks on a two-dimensional plane is a long open problem in the area of computational geometry, and it has an interesting history [DO07, Chap. 6]. Just to conclude, it has been proved that any linkage never gets locked on a two-dimensional plane (see [DO07, Sect. 12.1] for details). In other words, any linkage can unwind and extend to a straight line. Therefore, given the two folded states P and Q of the same linkage, both can unfold to the linear state

S. As with Theorem 5.2.3, this operation can be reversed at any time. If both P and Q can be unfolded to S, joining the operations of unfolding P to S and the reverse of the unfolding Q to S together, we can always transition from P to Q (and vice versa). □

Using the currently known constitutive proof of the theorem for "unlocking linkage", the operation "unfolding P to S" as described above corresponds to an operation that opens smoothly little by little the folded paper strip to the initial flat state. Namely, if you unfold each folded crease smoothly from the folded state of Fig. 5.4(2), you can always return to the state of Fig. 5.4(1). It is now proved that this is theoretically possible. It may seem intuitively trivial, but it took a long time to prove mathematically and strictly.

Considering the background, it is necessary to think carefully about the ability of simple folding. Here note that Theorem 5.1.1 only says that there exists at least one folded state for any given assignment. Also, by Theorem 5.1.2, there are many folded states other than pleat folding in general. Then, given a crease pattern string s and the folded state P that realizes s, can P be folded just by simple folding? From the example of Fig. 5.4, when the creases are not equally spaced, there exists a folded state where it cannot be folded by simple folding. Interestingly enough, however, in the case of equally spaced creases, any folded state can be reached just by simple folding. This proof is algorithmic and interesting that can give actual procedure as well.

Theorem 5.2.5 *Let P be the arbitrary folded state folded into unit length of the paper strip of length $n + 1$. Then, P can reach from the flat unfolded state S by simple folding. Furthermore, the number of simple folding can be bounded above by $2n$.*

Proof By Theorem 5.2.3, it is sufficient to show that from any folded state P, it can be extended to the initial state S by simple folding. Generally, in the folded state P, both endpoints are folded inside and invisible from the outside. Therefore, we consider dividing the operations of unfolding from P to S into two stages. That is, first, we consider making both endpoints of the paper visible from the outside, then extending it linearly. In order to analyze in detail, let p be the endpoint placed at the integer point $n + 1$ in the initial state S. Also, let us abuse P to denote "the current folded state", that is, P first indicates the given folded state, and it will be eventually changed to S after unfolding. Imagine when you finally obtain $P = S$ and represent each crease point on P by the value of the x coordinate in S. In other words, each crease point is represented by n integer points from 1 to n, and both endpoints are represented by integer points 0 and $n + 1$. Here we define *visibility* of a point on P in the folded state P. A point p is visible when p is placed on the surface of the folded state P. In Fig. 5.5, the visible points are represented by bold lines. We note that even if a crease point itself is visible, both line segments connected to this crease point may not be visible. For example, the crease point q in Fig. 5.5a is such a visible point. Such a crease point is a visible point of length 0. Here, we consider two cases separately depending on whether the endpoint p is visible or not. (In the context of the algorithm, these two cases correspond to two phases. That is, the algorithm first advances the processing of the first phase while the endpoint p is not visible, and proceeds to the second phase after p becomes visible.)

Case 1: The endpoint p is not visible in the folded state P (Fig. 5.5a–c are in this case). Let q be the visible point closest to p. That is, all points $r > q$ (including p) are not visible. Note that q may also be a point on a flat line segment. Now, among the points between p and q, let q' be the folded crease point closest to q with $q \neq q'$. When such a point does not exist (i.e., it is a line segment from q to p), we define $q' = p$.

First, consider the case when q is a point on a flat line segment. Since q is visible, the paper layers on the visible side of q on the side closer to q' can be folded by a simple folding at the crease point q' to the opposite side (when q' is a crease point, it can be opened by a simple folding). By this operation, the visible point closest to p changes from q to q'. Here, since q' is a point properly closer to p than q, the visible point approaches p.

Next, consider the case when q is a folded crease point. Without loss of generality, we can assume that the crease $q + 1$ is placed on the left of the crease q, as shown in Fig. 5.5a. Then, the paper layers placed on the opposite side of the point $q - 1$ with the segment q (which is the interval $[q, q + 1]$) in between are covering the non-visible point $q + 1$; however, since q is visible, they are not covering the point $q - 1$. That is, these paper layers can be turned over by a simple folding at the point q' closer to p than q using the crease point q'. (In Fig. 5.5a–c, the lower paper layers are flipped by a simple folding at the crease point r. This point r covers the crease point q', q' is closer to p than q, q' is the crease point, and it is not visible until the paper layers are flipped at the crease point r.)

In either case, the visible crease closest to p is updated from q to q' with $q' > q$ by this simple folding (which can be in fact unfolding by the simple folding). Continue

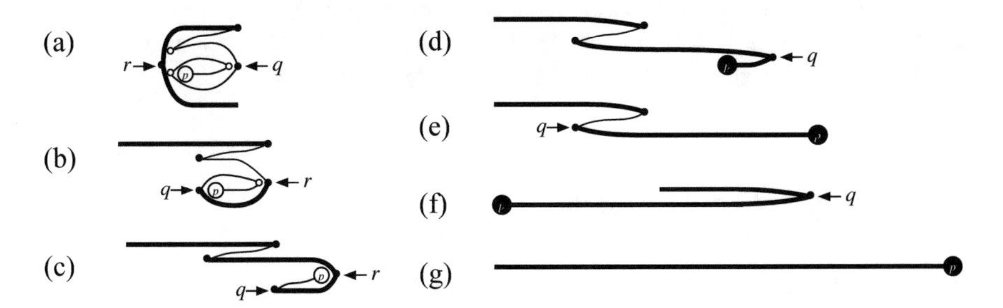

Fig. 5.5 An example of the sequence of simple (un)folding. The bold lines are visible from the outside, and the other lines are not visible. Each folded crease is indicated by • and ◦, where • is visible and ◦ is not visible from the outside. Note that two invisible lines can be joined by a visible crease point. The endpoint p is indicated by a large circle. In the algorithm, the first phase deals with Case 1, where the endpoint p is not visible from the outside, corresponding to (a)–(c), and the second phase deals with Case 2, where the endpoint p is visible from outside, corresponding to (d)–(g)

this procedure until the point p becomes visible. Since the coordinate is updated by at least one every time, the number of iterations is at most n. That is, the total number of folding (or unfolding) in case 1 is at most n, and eventually, it always transits to Case 2.

Case 2: The endpoint p is visible in the folded state P (Fig. 5.5d–f in this case). Let q be the closest folded point to the point p. If the point q is not visible, there is no folded point between the points p and q; since the point p is visible, the same argument as Case 1 is applied to the point q, and we can make q visible by simple folding. Thus, we can assume that all the points in the interval $[q, p]$ are visible. At this time, furthermore, these points are visible from the same side. (For example, assume that the interval $[q, r]$ is visible from the top, $[r, p]$ is visible from the bottom and not visible from the opposite side, respectively. Then, since p is the endpoint of the paper strip, it is not connected, which is a contradiction). Therefore, you can apply simple folding at the folded crease point q and flatten the paper at this point q. This operation does not affect the property that p is visible. Therefore, we can repeat this operation until all folding points are flattened. These two folding operations (folding that makes q visible if necessary, and that makes q flat) can actually be done all at once. Therefore, the total number of iterations of the simple folding is at most n.

From the above discussion, we obtain Theorem 5.2.5. □

Since there are slightly subtle points, we briefly summarize the results on one-dimensional equally spaced origami. If we have an M/V pattern at equal intervals on a paper strip, the following holds:

- Any pattern can be folded into the unit length (Theorem 5.1.1).
- Given any folded state, whatever it is, it can be folded by simple folding (Theorem 5.2.5).
- When you find out the procedure from the folded state P to unfold flat state S, you can also know how to fold P from S (Theorem 5.2.3).
- The pleat pattern, which is given by $MVMV\cdots$ or $VMVM\cdots$, has only one folded state, and all other patterns have multiple folded states.

5.3 Stamp Folding Problem

We first consider the number of ways of paper folding of length $n+1$. This is a classic open problem known by the name of *stamp folding problem*. To be exact, the stamp folding problem generally refers to the following problem.

Open Problem 5.3.1 *Fold the paper strip of length $n+1$ with n creases at equally spaced into unit length. Then how many folded states are there?*

That is, the stamp folding problem asks the number of ways of folding a paper strip of length $n+1$ into 1 regardless of MV assignments. Let $F(n)$ denote the number

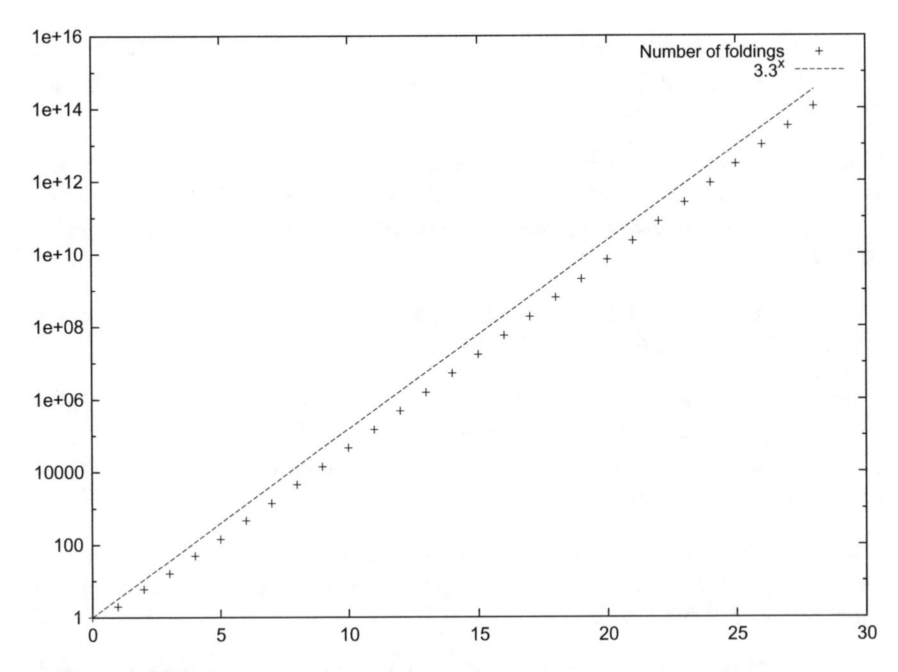

Fig. 5.6 Experimental results of the number of folded states. Each + mark is the actual number, and the dotted line is a graph of the function $f(x) = 3.3^x$

of ways of folding of the paper strip of length $n + 1$, which gives the answer to the stamp folding problem. Then while the exact value of $F(n)$ is unknown, the upper and lower bounds are investigated. The best upper and lower bounds known at the present are the theorem below[1]:

Theorem 5.3.1 *(1) The upper bound is* $F(n) = O(4^n)$. *(In other words, it is bounded above by* 4^n *with constant factor.) (2) The lower bound is* $F(n) = \Omega(3.06^n)$. *(In other words, it is bounded below by* 3.06^n *with constant factor.)*

Before proofs, we consider experimental values. A database that stores surprisingly huge integer sequences exists in the world: "The On-Line Encyclopedia of Integer Sequences".[2] When I entered the first few terms obtained in a simple program, I found that this sequence is recorded as A000136.[3] The data is recorded as "Number of ways of folding a strip on n labeled stamps", and it is exactly a sequence of stamp folding problem. Figure 5.6 is a plot of the function $f(x) = 3.3^x$ together with the sequence by +. As far as this graph implies, it seems to be considered experimentally that $F(n) = \Theta(3.3^n)$.

[1]Notations such as $O(f(n))$, $\Omega(f(n))$, $\Theta(f(n))$ are collectively called O-notation. Although they are not described in detail in this book, they are notations for bounding from above or below by using the main term of functions.

[2]https://oeis.org.

[3]https://oeis.org/A000136.

In the case of the general stamp folding problem, we do not mind MV assignments; however, looking especially at Theorem 5.1.2 on pleat folding, the following open problem is also interesting.

Open Problem 5.3.2 *Suppose that you fold a paper strip of length $n + 1$ at equally spaced for a given MV assignment into unit length. Then, what kind of MV string s that gives the most ways of folding? How many ways of folding are there for the MV string?*

We do not know the solution to this open problem now; however, from Theorem 5.3.1, the following corollary is obtained.

Corollary 5.3.2 *Let s be an MV string of length n generated uniformly at random. Then the expected value $f(n)$ of the ways of folding consistent with s is theoretically $f(n) = O(2^n)$ in the upper bound, $f(n) = \Omega(1.53^n)$ in the lower bound, and $f(n) = \Theta(1.65^n)$ experimentally.*

Proof The number of random MV string of length n is 2^n in total. Therefore, by the linearity of the expected value, each expected value is obtained by dividing each value of Theorem 5.3.1 by 2^n. □

By Corollary 5.3.2, for a string generated randomly, it takes exponential time if we enumerate all ways of folding that do not contradict the string. Let us calculate how long it takes. For example, suppose that the CPU of your computer is running at 5 GHz. Assume that a wise program on this computer finds one way of folding in just one clock. On the other hand, if you randomly generate an MV assignment of length 100, it has 1.53^{100} ways of folding on average. Then, this wise program finds all folded states in $1.53^{100} \times \frac{1}{5 \times 10^9} = 5.89 \times 10^8$ s. This is roughly 1120 years! This is somewhat too long.

Anyway, let me introduce how to prove upper and lower bounds below.

5.3.1 *Proof of Upper Bound*

We first show the upper bound.

Lemma 5.3.3 $F(n) = O(4^n)$.

Proof First, consider the case where n is an even number and let $n = 2k$ for some positive integer k. Suppose that the paper is folded and then placed in the interval $[0, 1]$. When looking at this paper layers from above, $n + 1$ segments are stacked in some order. Since there are $(n + 1)!$ different ways for ordering these $n + 1$ segments, we have a simple upper bound $(n + 1)!$. However, this is not a good upper bound. For example, if the paper layers come 1, 3, 2, and 4 in this order from the top, the paper ends up intersecting itself somewhere (Fig. 5.7). Taking this property into account, we give a better upper bound. Let us consider the vertical relationship of the paper

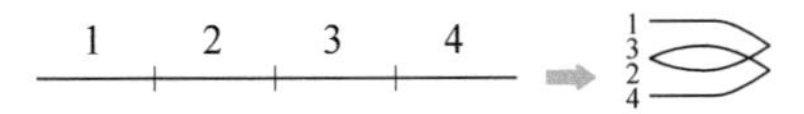

Fig. 5.7 If the paper layers are in the order 1, 3, 2, and 4, the paper penetrates somewhere

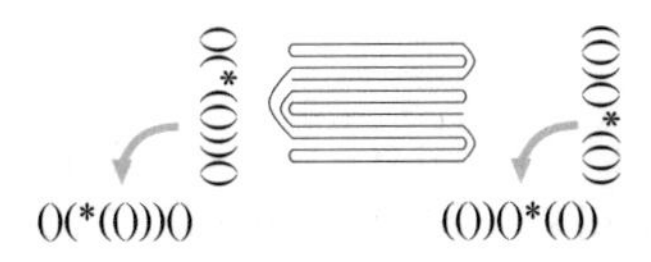

Fig. 5.8 Nest structures of a folded state: We have "$()(*(()))()$" on the left side and "$(())() * (())$" on the right side of the folded state

at coordinate 0. The paper must not penetrate each other means that k-folded crease points and one endpoint of the paper at point 0 (the left end of the segment 0) should be nested. Such nested structures have the same properties as "balanced brackets" and are studied very well. (See Fig. 5.8 for a simple example.)

Specifically, the number of nested structures of k pairs can be represented by the kth *Catalan number* C_k, which is defined by $C_k = \frac{1}{k+1}\binom{2k}{k} = \frac{(2k)!}{(k+1)!k!}$ (see, e.g., [Sta99]). Moreover, there are $(2k + 1)$ places in total to put the left end of the paper between these k pairs, which is indicated by $*$ in Fig. 5.8. Therefore, the number in the case of the left side paper connection at point 0 is at most $(2k + 1)C_k$.

Since exactly the same argument holds on the right side, the number in the case of the paper connection on the right side can also be bounded above by $(2k + 1)C_k$. Therefore, it can be bounded by $((2k + 1)C_k)^2$ in total.

Next, we consider the case that n is an odd number and $n = 2k + 1$ for some positive integer k. Since both ends of the paper come to the left side, the upper bound of the number on the left side can be bounded above by $(2k + 2)(2k + 1)C_k$ and the upper bound of the number on the right side is C_k. Therefore, the total upper bound is bounded by $(2k + 2)(2k + 1)(C_k)^2$.

Since $C_k \sim \frac{4^k}{k^{3/2}\pi}$ and $(4^k)^2 = O(4^n)$, the total upper bound is $O(4^n)$ in both cases.

□

At first glance, this upper bound seems to exactly check each folded state and count accurate values, not merely upper bounds, except the evaluation at the endpoints of the paper. In this discussion, however, the condition that "paper should be connected" is not considered at all. Therefore, it is not an exact value because the case that both sides are nested and the paper strip is disconnected is also counted. In order to bring the upper bound closer to the precise value, it is necessary to enumerate with considering "connectivity".

5.3.2 *Proof of Lower Bound*

Next, we show the lower bound.

Lemma 5.3.4 $F(n) = \Omega(3.065^n)$.

Proof We are given a paper strip with n creases of length $n + 1$. Here, we fix an appropriate value $k \ll n$ and consider folding the last k creases before the other creases. That is, we first fold this part into the unit length and glue it, then we consider a smaller problem with $n - k$ creases of length $n - k + 1$.

First, consider the folding of the part of this paper strip of length k. We let $G(k)$ denote the number of folded states of this part. That is, $G(k)$ represents how many ways of folding of paper strip of length $k + 1$ with k creases, but the left endpoint of the paper strip is not covered by the other folded paper. In other words, after folding, the leftmost edge of the paper is still visible from the outside, and it must be able to be attached paper of length $n - k$ on the outside. By repeating this procedure, you can obtain a lower bound of $F(n)$ by this function $G(k)$. Specifically, since we repeat $G(k)$ different ways of folding n/k times, $F(n) > (G(k))^{\frac{n}{k}} = (G(k)^{\frac{1}{k}})^n$ is established. Here, since k is a fixed integer, let us compute concrete values of $G(k)$ for some k. For example, it is not so difficult to compute $G(1) = 2$, $G(2) = 4$, $G(3) = 10$, $G(4) = 24$, and $G(5) = 66$.

Examining this sequence obtained in this way with "The On-Line Encyclopedia of Integer Sequences", it is registered as a sequence having an ID of A000682. (Incidentally, at this site, this sequence is registered as "Semi-meanders: number of ways a semi-infinite directed curve can cross a straight line n times". It is not straightforward whether this explanation fits with $G(k)$ or not. However, thinking that "a semi-infinite directed curve" is a paper strip of length n, and "a straight line" is a perpendicular line on the point $(n - k + \frac{1}{2})$, i.e., the central point of the interval $[n - k, n - k + 1]$, where the last paper strip of length k is folded into unit length and placed on the interval, the explanation of the semi-meanders exactly represents the function $G(k)$.) Since this function $G(k)$ is a monotone increasing function on k, the best result is obtained by using the largest value registered in the OEIS. Since the maximum value is $G(43) = 830776205506531894760$ so far (with offset 2), using this, the lower bound $F(n)$ for sufficiently large n is $F(n) > (830776205506531894760^{\frac{1}{43}})^n = 3.06549^n$. □

As already seen in Theorem 5.1.2, there is only one way to fold in the pattern of pleat folding. On the other hand, as you also saw in Example 5.1.1, even when you make a small change of the pattern of pleat folding, the way of folding increases exponentially. Moreover, as shown in Corollary 5.3.2, when we randomly generate an MV assignment, the way of folding is experimentally $\Theta(1.65^n)$; there are exponential combinations. Nonetheless, as we pointed out in Open Problem 5.3.2, we do not know what kind of patterns can be folded in the most ways. This is quite confusing to understand. Let us consider how many ways of folding actually exist for some patterns.

Exercise 5.3.1 Suppose you fold a paper strip of length $n + 1$ with n creases equally spaced into unit length. The same pattern repetition is described by the numbers on the shoulders. That is, $MMMM = M^4$, $MVMVMVMV = (MV)^3$, and so on. Furthermore, assume that n is an even number $n = 2k$ for some natural number k. Then consider how many ways of folding exist for the following patterns:

1. When all creases are in the same direction: M^n
2. When it differs from the last pattern of pleat folding: $(MV)^{n-1}MM$

Considering these concrete patterns, it seems difficult to determine quickly from a given pattern whether ways of folding are more or less.

Open Problem 5.3.3 *Suppose you fold a paper strip of length* $n + 1$ *with* n *creases equally spaced into unit length. Develop an algorithm that computes the number of ways of folding efficiently for a given crease pattern* s.

For example, if there is an algorithm that efficiently determines whether it becomes a polynomial or an exponential function with respect to n, it will be useful from the practical viewpoint.

References

[ABD+04] E.M. Arkin, M.A. Bender, E.D. Demaine, M.L. Demaine, J.S.B. Mitchell, S. Sethia, S.S. Skiena, When can you fold a map? *Computational Geometry: Theory and Applications* **29**(1), 23–46 (2004)

[DO07] E.D. Demaine, J. O'Rourke, *Geometric Folding Algorithms: Linkages, Origami, Polyhedra* (Cambridge University Press, Cambridge, 2007)

[Sta99] R.P. Stanley, *Enumerative Combinatorics*, vol. 2 (Cambridge University Press, Cambridge, 1999)

Chapter 6
Computational Complexity of Stamp Folding

Abstract In this chapter, we introduce two new notions of computational origami. The first one is "folding complexity", which is introduced to measure the number of folding. When you are given an origami design, you consider it is hard when the number of folding is more than one hundred. On the other hand, you feel it is easy when you obtain it after less than 10 times of folding. This intuition is formalized as folding complexity. The second one is "crease width". When you fold an origami model, if you have many paper layers at a crease, it is hard to fold them accurately. This intuition is formalized as crease width. We give some algorithmic results and hardness proofs about these new concepts.

6.1 Folding Complexity of Stamp Folding

In this section, we consider the notion of "folding complexity" and related problems. Specifically, we separate from the stamp folding problem and consider the following problem:

Input: A piece of paper strip of length $n + 1$ and a string s of length n over the alphabet $\{M, V\}$.
Output: The piece of paper strip with creases consistent with s.
Goal: Give a procedure with as few folding times as possible.

As a background of such problems, it is worth pointing out the relationship with the complex origami which is popular recently. Recent origami designers sometimes express complicated models in their crease patterns. In origami magazines, there is an expression method called "crease pattern folding", and experts in origami can fold the model out of a given crease pattern drawn on origami, without instruction. Sometimes it is not easy to divide the procedure into steps one by one, such as complicated, and when there are many lines to be folded at the same time. In such a case, it is sometimes difficult to draw the folding procedure. When doing such

R. Uehara, *Introduction to Computational Origami*,
https://doi.org/10.1007/978-981-15-4470-5_6

"crease pattern folding", they use the technique called "pre-creasing". This term means "folding in advance", in short, it is a method of preliminary folding crease lines on paper and folding the paper along these crease lines afterward. For example, there is an interesting video "How to Fold the MIT Logo in Origami in 3 Easy Steps".[1] In this video, Dr. Brian Chan creates a complex logo (Mens et Manus) of Massachusetts Institute of Technology from a single square piece of origami. It takes 3 hours to pre-crease out of the whole process, 3 hours to make the whole shape after that, and 4 hours to finish folding.

Such "pre-creasing" work is a monotonous and simple work; however, it is also an essential work for folding complicated origami accurately. How can we do these simple tasks efficiently? Because it is paper, you can pile the paper layers and fold them at once. Then, how can we fold and pile paper layers to make desired creases efficiently? This is the theme of this section.

Overlap and thickness of paper

Some readers may worry that accuracy will be ruined if you pile a lot of sheets of paper and fold them once. That problem is the theme in the next section. In this section, ignoring the error, we assume that the paper thickness is 0. In the society of computer science, it is well known that there is a trade-off relationship between time and space (or memory) required for computation. In origami, I think that folding efficiency and accuracy of folding are in a trade-off relationship. Research on such models themselves is also one of the interesting themes in the field of computation origami. Computational origami is a research field in the early days, so various models and problems are suggested, and intensified in the course of time, and valid models and problems will survive. This is exactly the discipline that is developing now.

When we count the number of folds, it is necessary to clarify the model of folding paper. Here we assume the "simple folding model" discussed in Sect. 5.2. That is, the paper is always in a flat folded state, once the crease line is confirmed, some layers of paper are valley folded. This operation is defined as "folding once". This number of times of folding will be minimized, and it is necessary to clarify two more things about that.

How to open folded paper: In order to make creases efficiently, it is necessary to open the folded paper and fold along another crease line many times. In this case, we need to think about the cost of an open operation of the paper. In this section, we simply assume that we can "completely open the paper and return it to a flat initial state" without any cost. In other words, when we perform "opening a paper", we will completely unfold it and return it to a flat initial state in one unfold operation with no cost.

[1] http://video.mit.edu/watch/the-making-of-mens-et-menus-in-origami-vol-1-2694/.

In fact, we should point out that this is not an essential difference. First, even when counting the number of times to unfold the paper one by one, it can actually open only the number of folds (by the same reason as Theorem 5.2.3). Therefore, even if the number of opening times is counted up one by one, it can be seen that the total number of operations can be bounded above by twice of the number of folds. When we consider algorithms, in general, such a constant factor is not considered very often. Thinking only of the number of folds will make the discussion much simpler; therefore, we follow this style in this book.

However, we should note that sometimes it is possible to save the number of folds by unfolding halfway and then executing the other folding. As far as I know, there is no research that evaluated the folding complexity of such fine folding.

Open Problem 6.1.1 *Construct a model that takes into consideration both the number of folds and the number of unfolds, with more elaborate folding computations than one in this section. Furthermore, design an efficient algorithm based on the model.*

How to make creases: We here clarify the criteria for making creases. We assume that each paper crease memorizes the last folded direction. Given the nature of paper, this definition has relevance and it is reasonable. That is, no matter which point on the given paper is taken, if the direction that was last folded is the direction we desire after folding, it is a reasonable folding method, and the one with the least number of folding times is the optimal folding algorithm.

6.1.1 Basic Properties of Folding Complexity

We first introduce two basic properties related to folding complexity.

Theorem 6.1.1 *To make n creases on a paper strip of length $n+1$, at least ($\lceil \log(n+1) \rceil + 1$) times of simple folding are necessary.*

Proof We first consider the case that the paper length is 2^k for some natural number k. That is, we have $n+1=2^k$. If you fold it somewhere, you will find that a part of the paper is doubled and the other part is still single. Considering putting as many creases as possible, it is better to fold the part that is double. Then, each point on the paper, the number of overlap is four or less. Likewise, it is the most efficient way to fold at the thickest part to make many creases. In short, it is best to fold the paper in half in the thickest part and make a crease as many layers as possible each time. Then after k folding, the paper thickness is 2^k, and there are $2^k-1=n$ creases. When the paper length n cannot be expressed as natural number k and $n+1=2^k$, take the smallest k where 2^k first exceeds $n+1$. Then we assume that the paper strip has length 2^k-1 (with extra paper strip) and do the same argument, the theorem can be obtained. □

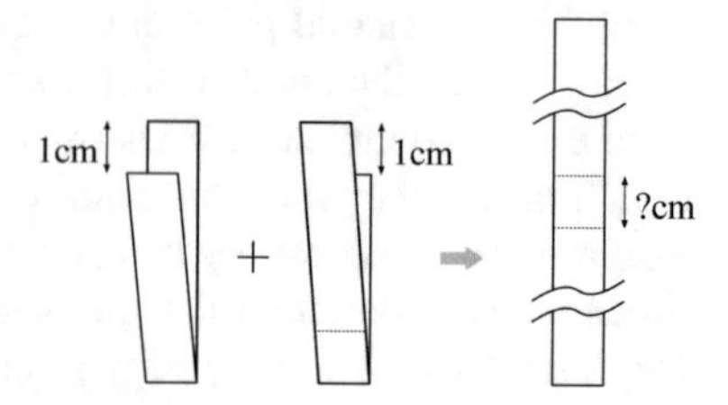

Fig. 6.1 Folding quiz

Theorem 6.1.2 *When we pile a paper strip by simple folding and fold it again, the creases which can be folded with the crease i are $i \pm 2, i \pm 4, i \pm 6, \ldots$. That is, the creases with different parities cannot be folded at once.*

Proof You can figure out as soon as you try; to overlap creases with different parities, you have to fold at point $i' + 1/2$ for some positive integer i'. Specifically, to put the crease i on the crease j, we have to fold at point $(i + j)/2$, which is not an integer if the parities of i and j are different. □

Theorem 6.1.2 is sometimes used as a puzzle; thus, it is a somewhat interesting property different from intuition. For example, solve the following quiz in your head and try it for real in practice. Can you obtain the answer correctly?

Exercise 6.1.1 Following Fig. 6.1a, fold the paper strip with 1cm shift, then fold the paper strip again with 1cm shift as shown in Fig. 6.1b. What is the distance between these two creases?

Considering such parity, the following corollary is particularly important in this section.

Corollary 6.1.3 *When there is an algorithm A that makes pleat folding, there is an algorithm A′ which makes the pattern $MMMM\cdots$ with the same folding complexity as A, and vice versa.*

Proof By Theorem 6.1.2, creases of odd indices and creases of even indices can only be folded completely and independently. Therefore, in the pleat folding pattern $MVMVMVMV\cdots$, any crease to be in M and another one to be in V cannot be folded once. In other words, any folding can be divided into folding for odd creases and even creases. From the algorithm A, we construct A' as follows; when the algorithm A folds an even crease, A' simulates A as is, and when the algorithm A folds an odd crease, A' reverses the folding direction. It is easy to see that A folds the pleat pattern if and only if A' folds the mountain folding pattern $MMMMM\cdots$. □

6.1.2 Algorithms for Pleat Folding

In this section, we consider efficient algorithms for making "pleat folding", that is, folding $VMVM\cdots$. By Corollary 6.1.3, it is sufficient to consider efficient algorithms for folding "all mountains". Actually, this is much easier to consider; hence hereafter, we will consider folding "a string consisting of only M of length n". In other words, we consider the problem of placing n consecutive mountain folds on paper of length $n + 1$. With the method in Corollary 6.1.3, we can easily construct an algorithm that folds pleat folding at the same number of times as the algorithm that solves this problem. As we saw in Theorem 6.1.1, at least $\lceil \log(n + 1) \rceil$ folding is required to have n creases. On the other hand, as with the proof of Theorem 5.1.1, if we fold in order from the end, we can easily fold n mountain creases with n times of simple folding. Then can we fold pleats more efficiently and properly smaller than n? The answer to this question is "Yes".

In this section, we first introduce an interesting algorithm from the viewpoint of analysis. This algorithm establishes that the pleat folding can be folded properly fewer than n. This algorithm is not very efficient compared to what we will introduce later; however, there is an interesting feature that *Fibonacci number* comes out in the analysis. Next, we introduce an ultra-fast algorithm. Finally, we show the theoretical lower bound. This lower bound is proved by a method called *counting argument*, which is a fairly powerful method. It can be seen that the abovementioned ultra-fast algorithm and the lower bound by the counting method are quite close. In other words, this ultra-fast algorithm is an algorithm with little room for improvement.

6.1.2.1 Fast Algorithm for Pleat Folding

First, we introduce an algorithm for pleat folding in properly fewer times than n. The goal here is the following theorem.

Theorem 6.1.4 *There is an algorithm of folding complexity $O(n^{0.69424\cdots})$ which folds "all mountain folding pattern" under simple folding model.*

Before the proof of the theorem, we think about the crease patterns that are made by folding paper strip in half each time. Since it is troublesome if the paper length is incomplete, when folding paper in half at center point k times, we consider the number of creases that are equally bisected each time. Since this number is $1 + 2 + 4 + 8 + \cdots + 2^{k-1} = 2^k - 1$, we assume that the paper length is $n = 2^k$ and the number of creases is $2^k - 1$. After folding this paper in half in the same direction at the center each time, when we unfold it gently, the folded paper is in a strange shape. Fixing so that all creases of this paper are at 90° gives an interesting shape with a recursive structure (Fig. 6.2). This is called *Dragon curve*, which has been studied extensively since ancient times [Gar67].

This dragon curve has $2^k - 1$ creases, but the orientation of each crease seems to be irregular at first glance. Is it true? Considering the creases made by the first four foldings, they are as follows:

- The first fold makes a crease in the center, and its orientation is M.
- The creases at the second fold are at 1/4 point and 3/4 point, and their directions are MV.

Fig. 6.2 Examples of dragon curves

- The third creases are folded at 1/8, 3/8, 5/8, 7/8, and their orientations are $MVMV$.
- The creases at the fourth fold are at 1/16, 3/16, 5/16, 7/16, 9/16, 11/16, 13/16, 15/16, and their orientations are $MVMVMVMV$.

It is not easy to see in the beginning; however, if you look at the creases of the fourth, the regularity is obvious. It can be generalized in the following way:

> By the ith fold, the 2^{i-1} creases are made at points $1/2^i, 3/2^i, 5/2^i, \ldots, (2^i - 1)/2^i$, and their directions are $MVMVMV \cdots MV$.

Here are two points to pay attention to.

- Our goal is to make all of them into M.
- The 2^{i-1} creases $MVMVMV \cdots MV$ are evenly spaced, and half of them are already M.

That is, half of these 2^{i-1} creases are already of the objective, and the creases of V in the other half are evenly spaced. How can we solve these equally spaced V creases? Looking closely, we can see that the equally spaced V creases occupy from end to end on paper. Therefore, as with the proof of Theorem 6.1.2, by folding only these equally spaced creases, we can create these creases independently of the other creases. In other words, we need to mountain fold all these equally spaced $2^{i-1}/2 = 2^{i-2}$ creases. Since this is the same problem as the original problem with smaller size, we can repeat the algorithm recursively.

We turn to prove Theorem 6.1.4. First we consider the case where the paper length is $n = 2^k$, and the number of creases is $2^k - 1$. The algorithm adopted here uses the property of the dragon curve. Specifically, we use the following algorithm A:

1. Fold the paper strip in half k times and unfold.
2. For each $i = 2, 3, \ldots, k$, we have 2^{i-2} creases in valley folding. We recursively apply A to each equally spaced 2^{i-2} creases.

Although it is quite confusing which creases should be folded if humans try to fold paper according to this algorithm, the basic idea of the algorithm is simple.

We analyze the number of folding operations by this algorithm A. Let $T_A(k)$ denote the number of folding operations when the algorithm A folds paper of length $n = 2^k$. Then, from the description of the algorithm, we have

$$T_A(k) = T_A(k-2) + T_A(k-3) + \cdots + T_A(2) + T_A(1) + T_A(0).$$

Since it is difficult to solve this equation as it is, we take a look at $T_A(k-1)$,

$$\begin{aligned} T_A(k) &= T_A(k-2) + T_A(k-3) + \cdots + T_A(2) + T_A(1) + T_A(0) \\ T_A(k-1) &= \qquad\qquad\quad\; T_A(k-3) + \cdots + T_A(2) + T_A(1) + T_A(0). \end{aligned}$$

Now, subtracting both sides yields $T_A(k) - T_A(k-1) = T_A(k-2)$, or equivalently, $T_A(k) = T_A(k-1) + T_A(k-2)$. This is exactly the definition of the general term of Fibonacci numbers. The usual Fibonacci sequence starts with $F_0 = 0$, $F_1 = 1$, $F_2 = 1$, $F_3 = 2$, and the general term F_n is

$$F_n = \frac{1}{\sqrt{5}}\left\{\left(\frac{1+\sqrt{5}}{2}\right)^n - \left(\frac{1-\sqrt{5}}{2}\right)^n\right\} = \frac{\phi^n - (1-\phi)^n}{\sqrt{5}},$$

where $\phi = \frac{1+\sqrt{5}}{2} \sim 1.6180\ldots$ is a number called the *golden ratio*.[2] Considering the cases where k of $T_A(k)$ are small here, $T_A(0) = 1$ because one folding is required to make $1 = 2^0$ crease, and $T_A(1) = 2$ because two foldings are required to make $2 = 2^1$ creases. Therefore, we have $T_A(k) = F_{k+2}$. In the Fibonacci numbers, the term of $(1-\phi)^n = (-0.618\ldots)^n$ rapidly converges to 0 and it is much smaller than the term of $\phi^n = 1.618\ldots^n$. Thus, we can consider $F_n = O(\phi^n)$.

Returning to the original problem, the number of folding operations required to fold the paper of length $n = 2^k$ is $T_A(k) = F_{k+2}$. Using $k = \log n$, $F_n = O(\phi^n)$, and the conversion expression at the base of logarithm (i.e., $\log_a b = \log_c b / \log_c a$), we obtain

$$\phi^{k+2} = \phi^2\phi^{\log n} = \phi^2\phi^{\log_\phi n/\log_\phi 2} = \phi^2 n^{1/\log_\phi 2} = \phi^2 n^{\log\phi} = \phi^2 n^{0.69424\ldots}.$$

Therefore $T_A(k) = O(n^{0.69424\ldots})$.

We here consider the general case for n such that there is no natural number k satisfying $n = 2^k$. We can use the same technique in the proof of Theorem 6.1.1. In this case, there is a natural number k that satisfies $2^{k-1} < n < 2^k$ uniquely; thus, we consider the paper strip of length 2^k. That is, the above algorithm can be applied by thinking that extra paper of length $2^k - n$ is virtually added to the paper of the length n currently considered. Since $2^{k-1} < n < 2^k$, the extra paper length is shorter than n. Therefore, under the O-notation, the result does not change.

This completes the proof of Theorem 6.1.4.

---

[2]For the golden ratio or Fibonacci sequence, see [Gar08].

6.1.2.2 Ultra-Fast Algorithm for Pleat Folding

Next, we show an ultra-fast algorithm for folding pleats. The running time of this algorithm is quite close to the theoretical lower bound shown in the next section, and hence it can be said that it is an almost optimal algorithm with little room for improvement.

Theorem 6.1.5 *There is an algorithm of folding complexity at most $\frac{3}{2}\log^2 n$ which folds "all mountain folding pattern" under simple folding model.*

We consider a case that the paper length is $n = 2^k$ for some natural number k again. (The other case of irregular length can be dealt with in the same way in the proof of Theorem 6.1.4.) That is, when you repeatedly fold k times in the center of the paper, the paper will be of unit length. We here introduce some symbols to show the state of creases. As before, M and V are a mountain crease and a valley crease, and let x be "not yet creased". We denote the left edge and the right edge of the paper after folding by [and], respectively. After unfolding, these creases are denoted by $+$. Each of these creases should have its own state of M or V; however, they will be added new creases M from above by this algorithm. Also, we use superscript to indicate repetition. For example, M^5 indicates $MMMMM$. We are now ready to show the ultra-fast algorithm to make all mountain foldings. It consists of three steps as follows:

Step 1: Repeat the operation "fold the paper in half at the center" $k-3$ times. We obtain the pattern of "$[xxx]$" on paper of length 4. Then make a pattern "$[MMM]$" by three mountain folding. Unfolding this will result in a crease pattern of "$+MMM+VVV+MMM+VVV+MMM+VVV+\cdots$".

Step 2: Fold the paper so that all sequences "VVV", which are made in Step 1, are overlapped at the same place. Precisely, in each folded state, continue to fold the paper in half at the center crease of the pattern "MMM" as close as possible to the center of the folded state. We note that when we repeat this step $k-3$ times, all the sequence "VVV" of length 3 will overlap at the same place. Here, since each pattern "MMM" is not exactly in the center of each folded state, after the $(k-4)$th folding, the last one folding is done to make it of the length 8. After this operation, the paper becomes a paper of length 8 of the pattern $[M+VVV+M]$. Now, we fold five times at each crease point of V and $+$ to make mountain creases, and obtain the folded state in a crease pattern "$[MMMMMMM]$". After unfolding the paper, we obtain the pattern "$VM+MMMMMMM+MVVVVVVM+MMMMMMM+MVVVVVVM+M\cdots$".

Step 3: Repeat Step 2. To be exact, perform the following for $i = 2, 3, 4, \ldots, k-2$:

1. Overlap the repeating pattern of V, i.e., "$VVV\cdots V$" on the same place. For this purpose, we fold the paper in half at the center of the

pattern "$MMM \cdots M$" which is as close to the center as possible each time.
2. Mountain fold all the creases with V and $+$, and unfold all.

Then, when the case $i = k - 2$ is done, only one sequence of consecutive "V" is left. Thus we mountain fold it all and finish.

Here we consider Step 1 as the first case of $i = 1$ and organize the algorithm. The algorithm performs the following operations for each $i = 1, 2, \ldots, k - 2$:

(a) repeat folding in half at the center of consecutive M part $(k - i - 2 + 1) = (k - i - 1)$ times (or $k - i - 2$ times only when $i = 1, k - 2$),
(b) pile all consecutive V parts onto one place, and
(c) fix the creases labeled "V" or "$+$" by $2i + 1$ folding.

Because it is somewhat confusing, we give an example. The pattern after (a) for the case $i = 6$ is

$$[M^{31} + M^{15}VM^7VMMMVMVVVVVMVMMMVM^7VM^{15} + M^{31}].$$

Intuitively, in the pattern after each (a), Vs concentrate in the vicinity of the center, and the other Vs and $+$s become farther and farther apart.

When carefully counting the number of folding operations, we have $(\sum_i^{k-2}(k - i - 2 + 1) + 2i + 1) - 2 = \sum_i^{k-2}(k + i) - 2 = k(k - 2) + (k - 2)(k - 1)/2 - 2 < \frac{3}{2} \log^2 n$.

6.1.2.3 Theoretical Lower Bound for Pleat Folding

Here we show a theoretical lower bound for pleat folding, that is, a theoretical limit that cannot be improved any further. This lower bound is close to the folding complexity (or upper bound) shown in Theorem 6.1.5, and they differ only in a constant factor. That is, it is difficult to improve the algorithm of Theorem 6.1.5.

Theorem 6.1.6 *In the problem of making n consecutive mountain folds on the simple folding model, the lower bound of the folding complexity is* $\log^2 n/4 \log\log n - o(\log^2 n/4 \log\log n)$.

Proof We assume that there is an optimal algorithm for making n mountain creases for each n, and its folding complexity is $f(n)$. Then, by Theorem 6.1.5, we have $f(n) \leq 2\log^2 n$. If there is a pair that satisfies $f(n_1) > f(n_2)$ even when $n_1 < n_2$, we can improve $f(n_1)$ by the paper of length n_1 as of length n_2 which is a contradiction to the definition of $f(n)$. Therefore, $f(n)$ is a monotonically non-decreasing function.

The algorithm folds $f(n)$ times to make n creases. Therefore, considering the expected value, at least once, the algorithm folds at least $n/f(n)$ paper layers to make $n/f(n)$ creases by one folding. To pile $n/f(n)$ sheets of paper, by Theorem 6.1.1, at least

$$\log(n/f(n)) = \log n - \log f(n) > \log n - \log(2\log^2 n) = \log n - 2\log\log n - 1 \tag{6.1}$$

foldings are required. On the other hand, when the algorithm folds $n/f(n)$ layers at once, the $n/2f(n)$ creases become valley folds.

These $n/2f(n)$ valleys are inevitable unless you open the paper flat on the simple folding model. To fix these $n/2f(n)$ valleys to mountains, the algorithm should fold at least $f(n/2f(n))$ times after unfolding the paper. (We note that these valleys are not uniform any longer, but this does not decrease the number of foldings and hence the algorithm still has to fold at least $f(n/2f(n))$ times.) Because the function $f(n)$ is a non-decreasing function and $f(n) < 2\log^2 n)$, we have $f(n/2f(n)) > f(n/4\log^2 n)$. We can use the same argument as before; this algorithm folds at least $(n/2f(n))/f(n/4\log^2 n)$ paper layers by one folding. To pile these paper layers, by Theorem 6.1.1, at least

$$\log \frac{n/2f(n)}{f(n/4\log^2 n)} > \log \frac{n}{4\log^2 n \cdot 2(\log \frac{n}{4\log^2 n})^2} > \log n - 4\log\log n - 3 \tag{6.2}$$

foldings are required.

After this folding, the algorithm folds $n/2f(n)f(n/4\log^2 n)$ creases, and we obtain $n/2f(n)f(n/4\log^2 n)$ valleys again. We can apply the same argument, which implies additional $\log n - 6\log\log n - 5$ foldings are required.

We can repeat the argument, and we can observe that $\log n - 2i(\log\log n + 1) + 1$ foldings are required at the ith iteration. This iteration ends up when $\log n - 2i(\log\log n + 1) + 1 \leq 1$, or consequently, $i = \lceil \frac{\log n}{2\log\log n} \rceil$. Therefore

$$\sum_{i=1}^{\lceil \frac{\log n}{2\log\log n} \rceil} \log n - 2i(\log\log n + 1) + 1 \leq \frac{\log n}{2\log\log n}\left(\frac{\log n}{2} - \log\log n - \frac{\log n}{2\log\log n}\right),$$

which completes the proof of the theorem. □

6.1.3 Algorithm and Lower Bound for General Patterns

We now turn to consider not only pleat folding, but also general patterns. As we have already seen, when we consider a folding algorithm, it is important how to fold the same patterns at the same time by piling them; however, we must also take care how to fix the inversion patterns that you have to make at that time. By adjusting this problem properly, we can design an algorithm that can fold any general pattern with truly fewer foldings than n. Precisely, using the algorithm shown in this section, any pattern of length n can be folded with $O(n/\log n)$ folding complexity. On the other hand, using information theory, we can show the folding complexity for a random pattern. As shown later, it is necessary to fold $\Omega(n/\log n)$ times with high probability to fold a randomly generated pattern. These results mean that almost all patterns require $\Omega(n/\log n)$ foldings, and there is an algorithm that makes it with $O(n/\log n)$ foldings. In this context, it turns out that pleat folding is an exceptional pattern that can be folded at ultra-high speed.

The function $O(n/\log n)$ is also related to efficiency of compression programs for strings. The data compression algorithms developed by Lempel and Ziv compress most strings of length n to $O(n/\log n)$ [LZ77, LZ78], and our algorithm in this section is based on a similar idea.

6.1.3.1 Folding Algorithm for General Patterns

We here introduce a folding algorithm for any general MV pattern of length n. The idea for reducing the folding complexity is to divide this pattern into certain short units and fold creases for each unit. We call this unit *chunk*. More concretely, we first decide the appropriate length s, divide the MV pattern of length n into n/s chunks, and fold each chunk according to the pattern. We will discuss the length of each chunk later, but we suppose that it is odd. The idea is that if you decide s suitable for a given n and build an algorithm corresponding to it, you can save the number of foldings. This algorithm works for any general MV pattern; of course, we cannot take so big s compared to n. Here we assume that there are at most k kinds of chunks when we divide the pattern into chunks. That is, while there may be quite many chunks, each chunk is equal to one of k different types of strings. Let $f(n, k, s)$ be the folding complexity for these n, k, and s in the worst case. In other words, given a pattern of length n and deciding the length s for it, if there are k different patterns, the pattern can be obtained by at most $f(n, k, s)$ folding operations. Then the upper bound of $f(n, k, s)$ is obtained as follows.

Lemma 6.1.7 *For an MV pattern of length n, let s be the length of a chunk and k the number of types of the chunks. Then the folding complexity $f(n, k, s)$ can be bounded above by $f(n, k, s) \leq \frac{4n}{s} + ks \log n$.*

Proof Let c be a chunk, and suppose that c appears t times. If we fold the paper in zigzag, then we can pile all of these t chunks to overlap them. This is always possible since the length of any chunk is odd. To fold this zigzag, it is enough to fold $t-1$ times. (We note that these chunks appear irregularly in the MV pattern in general. Thus, we cannot save the number of foldings here.)

After folding in a zigzag to pile the chunks, it is necessary to fix the creases that have folded in the opposite direction to the pattern we want since they are folded in the zigzag. Here, there are two ways to fold in zigzag depending on whether we fold into mountain or valley at the beginning. Of these two ways, if we choose the one with less direction to fold backward, we can reduce the number of creases folded in the opposite direction to less than or equal to half. Therefore, the total number of foldings can be bounded above by $t - 1 + t/2 \leq 3t/2$ (that is, $t-1$ times to overlap and $t/2$ times to correct the creases of opposite direction).

To fold creases in the chunks themselves, it is sufficient to fold s times. However, because they are piled by zigzag folding, the half can be folded correctly, but the other half will be folded in the opposite direction. In order to fix this, the same processing is performed recursively for half of the opposite direction. Considering that it is halved each time, the number of folding in the whole can be bounded by the following formula:

$$\left(\frac{3t}{2}+s\right)+\left(\frac{3t}{4}+s\right)+\left(\frac{3t}{8}+s\right)+\cdots+(1+s)<3t+s\log\frac{3t}{2}.$$

We assume that the ith-type chunk appears t_i times in the original pattern for each i. That is, we have $\sum_{i=1}^{k} t_i=\frac{n}{s}$. When we repeat the above operation for each i of the k chunks, the folding complexity for folding all the creases is

$$\sum_{i=1}^{k}\left(3t_i+s\log\frac{3t_i}{2}\right)\leq 3\frac{n}{s}+ks\log n.$$

Finally, considering the creases at the seam between two consecutive chunks, since there are $\frac{n}{s}$ chunks, we add a procedure to fold each one individually, and we obtain

$$f(n,k,s)\leq\frac{4n}{s}+ks\log n.$$

□

By Lemma 6.1.7, the upper bound $\Theta\left(\frac{n}{\log n}\right)$ of the folding complexity for any pattern is immediately obtained.

Theorem 6.1.8 *The folding complexity of any pattern of length n is at most*

$$(4+\varepsilon)\frac{n}{\log n}+o\left(\frac{n}{\log n}\right)$$

for arbitrary $\varepsilon>0$.

Proof In Lemma 6.1.7, we let $s=(1-\epsilon)\log n$, $k=2^s=n^{1-\epsilon}$, and $\epsilon=\frac{\varepsilon}{4+\varepsilon}$. Then we have the following:

$$\frac{4n}{(1-\epsilon)\log n}+(1-\epsilon)n^{1-\epsilon}\log^2 n=(4+\varepsilon)\frac{n}{\log n}+o\left(\frac{n}{\log n}\right).$$

□

6.1.3.2 Lower Bounds for General Patterns

In this section, we show that most of the general patterns are hard in term of "folding complexity". To put it in more detail, we show that the folding complexity of a randomly generated MV pattern is $\Omega(n/\log n)$ with high probability. This complexity is equal to one achieved by the folding algorithm in the previous section within a constant factor. In other words, for most patterns, the folding algorithm in the previous section is essentially optimal, and there is little room for improvement except a constant factor. That is, patterns such as pleat folding and dragon curve are actually exceptional ones which admit us to fold efficiently.

Theorem 6.1.9 *The folding complexity of a random string of length n is at least*

$$\frac{n}{2+\log n}$$

with high probability.

Proof Here we prove the theorem by method called *counting argument*. We first estimate how many folded states there are when we fold a paper strip k times. Since the number of random strings of length n is 2^n, the number of folded states when we fold k times should be at least 2^n; otherwise, there exist strings that cannot be made by k foldings. This is the essence of the counting argument.

There is an option in which crease is to be folded; there are at most n creases. When folding each crease, there are also two choices: whether mountain folding or valley folding. Moreover, when we make k foldings, we have another option of "unfolding all the paper" between two foldings. This means that we are making k selections of either unfolding or not. (Considering only the "between" of two folding operations, we have $k-1$ selections, but considering the meaningless operation of unfolding all in the end after all foldings are complete, it may be considered as k selections.) Therefore, the number of the folded states after the "k folding" (with unfoldings sometimes) is at most $(n \times 2 \times 2)^k = (4n)^k$.

Therefore, if $(4n)^k = o(2^n)$, most patterns cannot be folded. In other words, the folding complexity for most patterns must be k that satisfies at least $(4n)^k > 2^n$. Solving $(4n)^k = 2^n$, we obtain $k = \frac{n}{2+\log n}$. □

We note again that the upper bound by Theorem 6.1.8 and the lower bound by Theorem 6.1.9 differ only by around four times. That is, for most of the MV patterns, the way of folding by the algorithm indicated by Theorem 6.1.8 is almost optimal, and there is little room for improvement. Since these upper and lower bounds depend on the folding model, in order to shrink the constant factor 4, it is necessary to further elaborate the model and make the discussions of both bounds into details.

Anyway, by Theorems 6.1.8 and 6.1.9, most patterns cannot be folded efficiently. That is, patterns such as pleat folding and dragon curve are extremely exceptional special patterns. However, Theorem 6.1.9 shows just existence based on the counting argument, and it does not give specific patterns that cannot be folded efficiently. That is, the following problem is open.

Open Problem 6.1.2 *Give a concrete MV pattern where the folding complexity is $\Omega\left(\frac{n}{\log n}\right)$.*

Of course, the following problem is also open.

Open Problem 6.1.3 *For a natural number k, what kind of concrete MV pattern whose folding complexity is $O(\log^k n)$ except pleat folding or dragon curve?*

For example, intuitively, if an MV pattern is close to pleat folding, its folding complexity would be small; however, how can we decide it is "close"?

As we saw in Theorem 5.1.2, pleat folding is the only one where crease pattern is regular and only one folded state exists. On the other hand, the MV pattern in Example 5.1.1 looks almost the same as the pleats; however, there are exponentially many folded states. What is the folding complexity of such a pattern? To be more specific, the following open problem is interesting.

Open Problem 6.1.4 *Is there any relationship between the folding complexity and the number of folded states?*

6.2 Crease Width of Stamp Folding

In this section, following the notion of "folding complexity", we introduce the other new notion of "crease width" and consider the problems related to it. Specifically, we consider the following problem:

Input: A paper strip of length $n + 1$ and a string s of length n over the alphabet $\{M, V\}$.
Output: The paper strip folded to unit length.
Objective: Find a folded state as little paper as possible in the crease.

As we thought in the first column in this chapter, in the field of computer science, the time required for computation and the amount of memory used for computation are considered as the cost of computation or computational resource. Of course, it is an excellent algorithm that the number of computation steps is small, and the memory consumption is low. What kind of things can be considered for such resources and costs in origami? First, the most natural idea would be "the number of foldings". It corresponds naturally to the time complexity. This is "folding complexity" that we consider in Sect. 6, and there would be little dispute about this. Then what is the counterpart of the amount of memory? There are many ways of thinking about this. The "crease width" introduced in this section is simply "the number of sheets of paper caught in a crease". This is an idea from the observation that "when we fold many sheets of paper at once, their creases are shifted". You can understand immediately when you actually fold it, but if you want to fold the origami precisely, it is better not to overlap too much.

In fact, in the case of making sophisticated origami works, it is common to first carefully fold only creases, and finally take the technique of three-dimensioning it along creases. If you overlap several sheets, sometimes more than ten sheets, it is difficult to make beautiful and precise creases. Considering this background, it is better to have fewer sheets of paper in creases. On the other hand, in order to reduce the folding complexity, it is necessary to fold many sheets of paper at once. This relationship is similar to a trade-off relationship between time complexity and space complexity in computer science such as "It takes a lot of memory when trying to compute it at high speed" or "It takes a long computation time to save memory".

For these reasons, we propose using the notion of "crease width" as an indicator of cost in origami. Of course there can be some objections on this, and it is also welcome that different indicators will be proposed in the future. However, in any rate, the problem of "reducing the crease width" is a very interesting research theme from the viewpoint of computer science, and it is a notion worth being investigated. We introduce its interesting properties in this section.

We first confirm that this problem is feasible for any given pattern s. By Theorem 5.1.1, we note that any pattern can be folded to unit length. Also, in any folded state, it is guaranteed that it can be folded by simple foldings by Theorem 5.2.5. Moreover, when a certain MV pattern s is given and s represents pleat folding, then it has a unique folded state by Theorem 5.1.2, there are no paper layers pinched in any crease, and the problem in this section can be solved instantaneously. The following corollary is immediately obtained by thinking a little.

Corollary 6.2.1 *The following two are equivalent: (1) The folded state is a pleat folding. (2) The number of paper layers at every crease is 0.*

That is, if the given pattern s gives pleat folding, the problem can be immediately solved and finished; however, in all other cases, some paper layers must be caught at some creases, there are two or more folded states, and hence there is a meaning of considering the "optimal" folded state among them.

6.2.1 Optimization Problem and Computational Complexity

So far, "crease width" has been vaguely defined as "the number of sheets of paper caught in a crease", but now we properly define the problem in detail. The input of the problem is a paper strip P of length $n + 1$ and string s of M and V of length n. The output is the paper strip P folded to unit length consistent to s. A feasible folded state is always present by the above consideration. If s indicates pleat folding, there are only one folded state; however, there are two or more folded states in other cases.

When a fold state of P is given, the *crease width at crease i* is defined by the number of paper layers caught at the crease i (Fig. 6.3). Then, we can consider the following three minimization problems for the crease width in the folded state of the paper P.

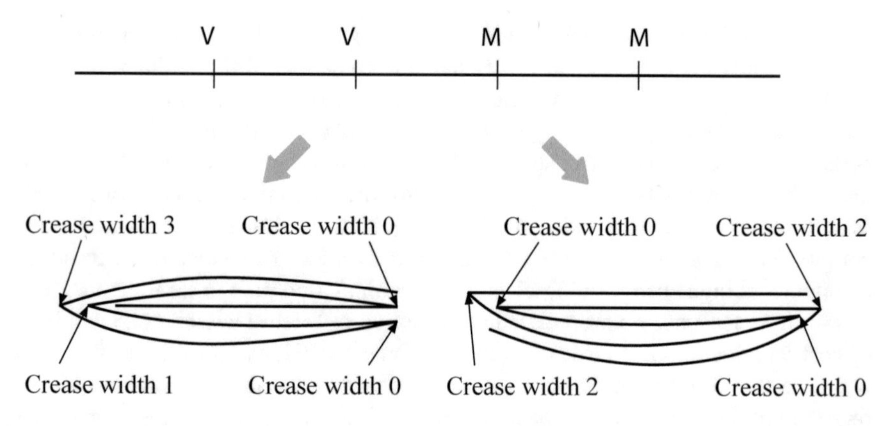

Fig. 6.3 Examples of the folded states for the string $VVMM$ and the crease width at each crease

Minimize the maximum value: Consider the crease width of each crease and minimize the maximum value.

Minimize the average value: Consider the crease width of each crease and minimize the average value.

Minimize the total value: Consider the crease width of each crease and minimize the total value.

Considering that the total value divided by n is the average value, it turns out that two of the three problems are essentially the same problem. Therefore, hereafter, we only consider the minimization problem of the maximum value and the minimization problem of the total value. Here we think about the difference between the two problems. At first glance, it is not even clear whether these two problems are different or not. Strictly speaking, even at the time of this writing, there is no clear difference. Here we consider the pattern of Example 5.1.3 again.

Example 6.2.2 Consider again the problem of folding a paper strip of length 12 consistent with $s = MMVMMVMVVVV$. There are 100 ways to fold this strip into unit length. Among these 100 ways, there is only one way of folding that minimizes the maximum crease width, there is only one way of folding that minimizes the total crease width, and these two ways of folding are different. Each way of folding is shown in Fig. 6.4.

As introduced in Example 5.1.3, there is no other known method but to try all the possible combinations to solve these problems so far. The above solutions are also the results of the exhaustive search by the program created by me. That is, my program first enumerated all the folded states consistent with this MV pattern, computed the crease width for each crease, and output the value. Intuitively, there seems to be a way to virtually fold creases of the paper according to the string s; however, if you try to make a program, this is pretty tough. In fact, Algorithm 4 is easier to implement, which I actually did.

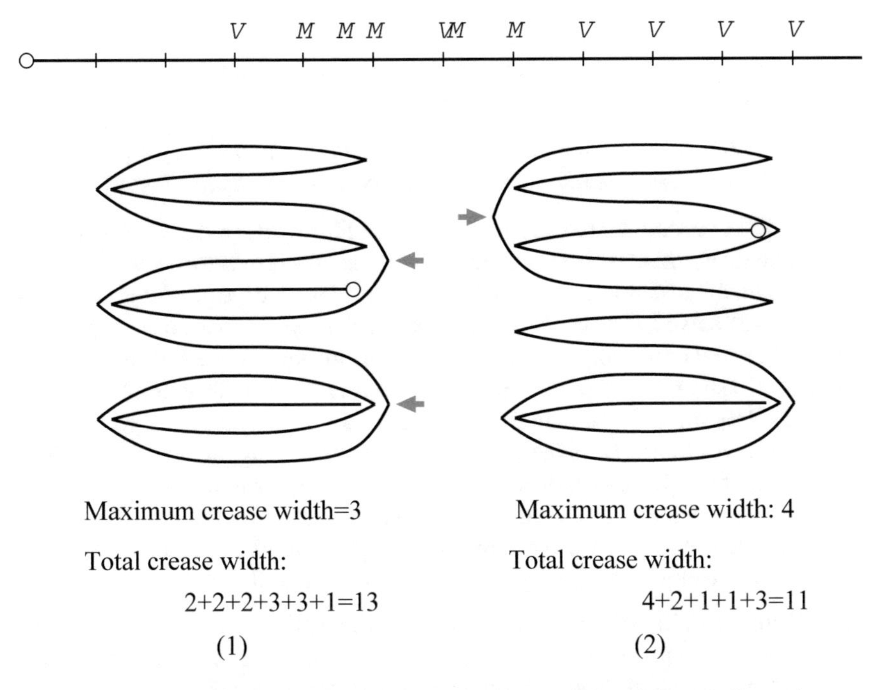

Fig. 6.4 Two folded states that minimize the crease width relative to the string $s = MMVMMVMVVVV$. Each crease with the largest crease width is indicated by an arrow. (1) The (unique) folded state in which the maximum crease width is the minimum value 3. (2) The (unique) folded state in which the total crease width is the minimum value 11.

Input : An MV string s of length n;
Output: All folded states consistent with s;
for *each permutation Q from 1 to n* **do**
  Suppose that Q is a folded state of a paper strip and check whether this realizes s;
  Check if both ends of the folded paper strip in unit length are nested properly;
  if *both conditions are satisfied* **then** output the folded state Q realizing s
end

Algorithm 4: An enumeration algorithm all folded states for the string s

In Algorithm 4, checking whether the segments of paper are nested properly at the two ends is the same as the argument in Lemma 5.3.3, and it is necessary to check whether the folded paper intersects itself or not. This is essentially the same as checking balanced brackets as appeared in the proof of Lemma 5.3.3, and it is easy to check if you use stack (in linear time). There are some known algorithms for enumerating all permutations per constant time. Therefore, the above algorithm runs in $O(n \cdot n!)$ time. This is an exponential function, and hence it is an inefficient algorithm from the theoretical viewpoint, but even with my naive program, it works in a realistic time up to $n = 20$ or so. The MV sequence $s = MMVMMVMVVVV$ introduced in Examples 5.1.3 and 6.2.2 was found by this program.

Then, is there any way to find such an answer smarter than this naive way? In this section, we will see that it does not seem to be likely. The first theorem claims that the problem of minimizing the maximum crease width is theoretically intractable.

Theorem 6.2.3 *For a given paper strip P of length $n + 1$ and an MV pattern s of length n, the problem of minimizing the maximum crease width is NP complete.*

For supplementing the readers who are unfamiliar with the theory of computational complexity, the class NP is a class of problems closely related to the millennium problem called P≠NP conjecture. (*Millennium problems* are the problems with prize money; if you solve it, you can be a millionaire.) The P≠NP conjecture is generally predicted to be P≠NP, where P is a class of problems that can be solved efficiently in a sense. In short, if you prove that a problem is an NP complete problem (if you believe P≠NP, which is the majority), which gives a strong basis that you will not be able to solve efficiently. In the context of our paper folding problem, the theorem claims that, after all, we cannot solve the crease width problem of minimizing the maximum value without trying out all combinations unless P=NP.

On the other hand, as for the minimization problem of total crease width, its computational complexity is not yet shown so far and is still open:

Open Problem 6.2.1 *When the paper strip P of length $n + 1$ and the MV pattern s of the length n on it are given, can the problem of minimizing the total crease width be solved in polynomial time? Or is it an NP complete problem?*

Computational complexity is not yet established, but a partial answer is obtained as follows.

Theorem 6.2.4 *When the paper strip P of length $n + 1$, the crease pattern s of the length n, and the natural number k are given, the problem of deciding whether or not the total crease width is less than k can be solved in $O((k + 1)^k n)$ time.*

In a term of the computational complexity, it is said that the problem of determining the total crease width is *fixed parameter tractable (FPT)*. Roughly speaking, the problem of deciding whether total crease width is less than k can be solved efficiently when k is small. That is, if k is a sufficiently small constant, since $(k + 1)^k$ is also constant theoretically, the algorithm of $O((k + 1)^k n)$ time runs in linear time, which is a fast algorithm. However, if k is as large as proportional to n, this algorithm takes exponential time, and then it does not differ from Algorithm 4 that checks all combinations.

Now we prove each of these two theorems.

6.2.2 NP Completeness of the Minimization Problem of the Maximum Crease Width

In this section, we prove that the problem of minimizing the maximum value of the crease width is NP complete. It is obvious that this problem is in NP, thus we prove

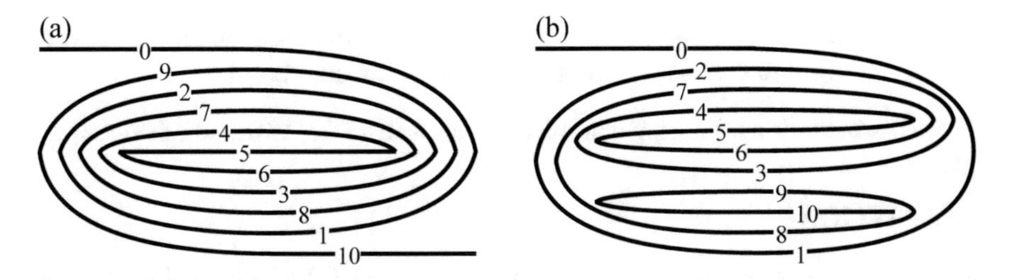

Fig. 6.5 **a** Spiral folded state and **b** not spiral folded state

its hardness. In order to prove NP hardness, it is sufficient to simulate some other problem which is already known to be NP hard with the current problem. This is called a *reduction* in a technical term.

Here, we remind that, by Theorem 5.1.2 and Corollary 6.2.1, the pleat folding is the unique pattern in the sense that it has only one folded state and every crease has crease width 0. We now first consider another pattern with a special property such that all creases are assigned to the same direction. For this special pattern, we have the following.

Observation 6.2.5 *Let s be an MV pattern V^n or M^n, i.e., a sequence of the same letter, for an integer n. Then there are n folded states for the pattern s, and the crease width in each folded state is $n-1$.*

This observation is obvious by Fig. 5.1. Combining two of these patterns gives us a useful pattern.

Observation 6.2.6 *For a natural number n, let s be an MV pattern M^nV^n. Then there are n^2 folded states for the pattern s. If the first segment 0 and the last segment $2n$ can be seen from the outside in the folded state, the folding state is either* $[0|2n-1|2|2n-3|\dots|2i|2(n-i)-1|\dots|1|2n]$ *or its reverse.*

Two folded states for the string $s = M^5V^5$ are shown in Fig. 6.5. In Fig. 6.5a, the 0th and 10th segments are visible and are in the folded state referred in Observation 6.2.6. We call this folded state a *spiral folded state* of length $2n$. Other folded states are ones where either the 0th and 1st segments are visible as shown in Fig. 6.5b, or the $(2n-1)$st and $(2n)$th segments are visible.

Now, we show the NP hardness of the next decision problem.[3]

Input: A paper strip of length $n+1$, an MV pattern s of length n, and a natural number k.

[3]Some readers may think that the difficulty of a problem is slightly different between the optimization problem and the decision problem. However, they are considered to have essentially the same difficulty in the following way. First, if the decision problem is solved, by finding the boundary between "Yes" and "No" of the decision problem for each $k = 1, 2, \dots, n$, we can solve the optimization problem. Also, if we solve the optimization problem, we can solve the decision problem by using the answer.

Output: Decide whether we can make the maximum crease width less than or equal to k when we fold the paper strip into unit length.

We use the following decision problem called *3-PARTITION* to show the hardness ([GJ79]):

3-PARTITION

Input A set $A = \{a_1, a_2, \ldots, a_{3m}\}$ with $3m$ elements and a natural number B. Here, each element a_i has a natural number weight $w(a_i)$, and each weight satisfies $B/4 < w(a_i) < B/2$ and $\sum_{j=1}^{3m} w(a_j) = mB$.

Output Decide whether A can be split into m disjoint subsets $A^{(1)}, A^{(2)}, \ldots, A^{(m)}$ such that $\sum_{a_j \in A^{(i)}} w(a_j) = B$ for each $1 \le i \le m$.

By condition $B/4 < w(a_i) < B/2$, we can assume that $\left|A^{(i)}\right| = 3$ for each i. It is well known that the 3-PARTITION problem is strong NP complete [GJ79]. Therefore, it is sufficient to show (a polynomial time) reduction from 3-PARTITION to the minimization problem of the maximum crease width. In other words, when one instance of the 3-PARTITION problem is given, we construct a paper strip P and an MV pattern s such that solving the minimization problem of the maximum crease width for P and s implies solving the instance of the 3-PARTITION. (Intuitively, if we can make such a reduction, we show that the problem of minimizing the maximum crease width has the same (or more) difficulty as 3-PARTITION.) Therefore, in the following, we assume that $a_1, \ldots, a_{3m}$ and B are given as inputs to 3-PARTITION, and show a construction of the inputs s and k of the minimization problem of the maximum crease width.

Construction of the MV pattern s and the natural number k: First, for each element $a_j \in A$, consider the next string x_j.

$$x_j = \begin{cases} V^{w(a_j)m^3} M^{w(a_j)m^3} & \text{if } j \text{ is odd} \\ M^{w(a_j)m^3} V^{w(a_j)m^3} & \text{if } j \text{ is even.} \end{cases} \tag{6.3}$$

Next, for each $j = 1, 2, \ldots, 3m$, the string s_j is defined as follows:

$$s_j = \begin{cases} (VM)^m x_j (VM)^m & \text{if } j \text{ is odd} \\ (MV)^m x_j (MV)^m & \text{if } j \text{ is even.} \end{cases} \tag{6.4}$$

Furthermore, we define two strings $t_1 = M^{2Bm^3+16m^2}$ and $t_2 = V^{Bm^3+8m^2+1} M^{Bm^3+8m^2}$. We call these two strings t_1 and t_2 *terminators*. Also, we let $f = (MV)^{m+1}$. Now we are ready. The desired MV string s is defined by joining all these substrings as follows:

$$s = t_1 f t_2 s_1 s_2 \ldots s_{3m}.$$

Then we let $k = 2Bm^3 + 16m^2$. The string s and the natural number k can be computed in polynomial time from the original instance of 3-PARTITION.

Let us first explain the intuitive intent of this string s. Since t_1 and t_2 should be in the thick spiral folded states, they will come to both ends in the final folded state. The string f between these terminators is a pleat folding, and the ith valley fold of this pleat folding represents the set $A^{(i)}$, and it aims to sandwich the other paper layers later. We call this ith valley fold of f as the ith *folder*. Each x_j sandwiched between them has the crease width corresponding to the element a_j of the original 3-PARTITION and aims to be pinched in each folder in equal crease width (that is to fit in k).

Let us show two lemmas. Combining these two lemmas, we can prove Theorem 6.2.3.

Lemma 6.2.7 *Suppose that an instance $\{a_1, a_2, \ldots, a_{3m}\}$ with B of 3-PARTITION has a solution. Then the MV pattern s constructed above has a folded state with the maximum crease width at most k.*

Proof Let $A^{(1)}, \ldots, A^{(m)}$ be a partition of A that gives a solution of 3-PARTITION. That is, $A = \dot{\bigcup}_{i=1}^{m} A^{(i)}$. Then, for each i with $1 \leq i \leq m$, we have $A^{(i)} \subset A$, $\left|A^{(i)}\right| = 3$, and $\sum_{a_j \in A^{(i)}} w(a_j) = B$. We construct a folded state of crease width k of one-dimensional origami from this partition.

We first consider the folded state of the partial string $s' = t_1 f t_2$. By Observation 6.2.5, the number of folded states of the terminator t_1 is $k + 1$. Among them, there is only one way of folding where "end" appears externally that can connect from t_1 to f. If this "end" of t_1 is folded inside, all paper layers in $f t_2$ are folded inside along with the edge of t_1, and the maximum crease width exceeds k. Therefore, if the maximum crease width is limited up to k, the folding way of t_1 should be $[k|k-2|\ldots|k-2i|\ldots|2|0|1|3|\cdots|2i+1|\ldots|k-1]$. Since the folding way of pleat folding f is uniquely determined, the folding state is $[k+1|k+2|\cdots|k+m+2]$. Next, consider another terminator t_2. In order to fold this part of crease width at most k, we can only pinch this part into $(m+1)$st folder with spiral folded state $[k+m+3|2k+m+3|k+m+5|2k+m+1|\cdots|2k+m+2|k+m+2]$.

So far, the substring s' can be folded as follows (Fig. 6.6):

$$[k|k-2|\cdots|k-2i|\cdots|2|0|1|3|\cdots|2i+1|\cdots|k-1|k+1|k+2|\cdots|k+m$$
$$|k+m+1|2k+m+3|k+m+3|2k+m+1|\cdots|2k+m+2|k+m+2].$$

For each $A^{(i)} = \{a_j, a_{j'}, a_{j''}\}$, intuitively, fold the corresponding substrings x_j, $x_{j'}$, $x_{j''}$ and put them together in the ith folder. The details are described as follows.

We first consider when j is odd. Let us fold the part of the substring x_j and place it in the ith folder. This folding consists of three steps (Fig. 6.7). The first $(VM)^m$ in s_j is folded as follows. Each M, or mountain fold, is placed one by one in descending order from the end of m folders. When it reaches the ith folder on the way, the remaining sequence of (VM)s is folded into the ith folder then.

The next substring x_j is folded in a spiral folding and placed in the ith folder. More precisely, the substring x_j is in the interval of $[t..t+2w(a_j)m^3+1]$ of the whole paper string. Then fold it into spiral folding $[t|t+2w(a_j)m^3-1|t+2|t+$

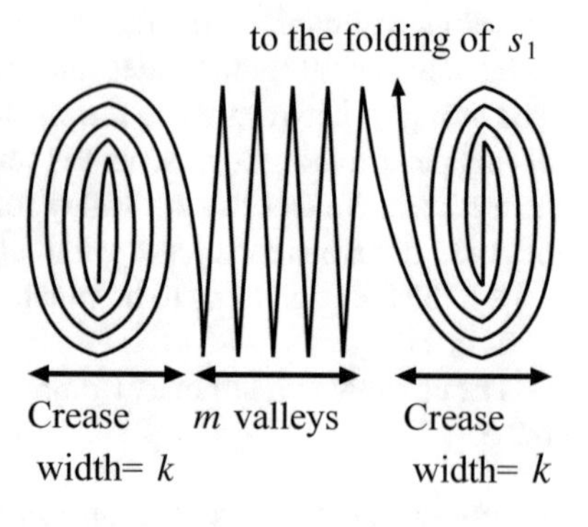

Fig. 6.6 The folded state after folding two terminators t_1, t_2, and pleat folding f between them

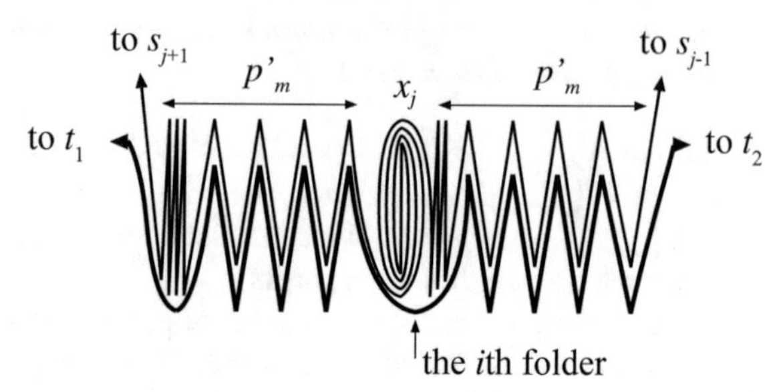

Fig. 6.7 How to fold the substring $s_j = (VM)^m x_j (VM)^m$

$2w(a_j)m^3 - 3|\cdots|t + 2l|t + 2(w(a_j)m^3 - l) - 1|\cdots|t + 1|t + 2w(a_j)m^3]$, and then store it in the ith folder.

For the second substring $(VM)^m$, each mountain fold is put into a folder in descending order one by one from the ith folder. When finally arriving at the first folder, all the remaining pleats are stored in this folder. We note that each substring $(VM)^m$ has enough number of zigzag, and hence we can put every pleat in each folder so that any folder is not shut.

When j is even, the substring s_j is stored in the same way as in the odd case, but in reverse as a whole, it is in ascending order instead of descending order. Intuitively, we will alternately fill leftward and rightward odd-numbered and even-numbered folded gadgets over pleat folding folders between the terminators.

We show that the maximum crease width of the finally obtained folded state does not exceed k in this folding way. In the folded part of the terminator t_1, the maximum crease width k is achieved between segments k and $k - 1$. Likewise, the maximum crease width in the fold of the terminator t_2 is k between $k + m + 3$ and $k + m + 2$. Next, in the ith folder, we have three spiral folding states of the three substrings $x_j, x_{j'}, x_{j''}$ and some pleat foldings corresponding to $(VM)^m$ or $(MV)^m$ of substrings x_h for some $h \in \{1, \ldots, 3m\} \setminus \{j, j', j''\}$. Therefore, in total, there are at most $2(w(a_j)m^3 + w(a_{j'})m^3 + w(a_{j''})m^3) + 2 \cdot 2 \cdot m \cdot 3m = 2Bm^3 + 12m^2$ paper layers, which is less than k by the definition of k. This completes the proof of Lemma 6.2.7. □

Next we show the opposite claim in the following lemma.

Lemma 6.2.8 *For the given MV pattern s, if a folded state with the maximum crease width k exists, the original instance $\{a_1, a_2, \ldots, a_{3m}\}$ with B of 3-PARTITION has a solution.*

Proof First, let us show that the following three conditions are satisfied in the folded state where the maximum crease width is (at most) k:

(1) The substring $s' = t_1 f t_2$ is in the unique folded state shown in Fig. 6.6.
(2) The substring x_j is folded into one folder.
(3) Each folder sandwiches the folded states corresponding to exactly three x_js.

By Observations 6.2.5 and 6.2.6, Condition (1) follows. Also, we can see that the maximum crease width already becomes k in this part. Condition (1) implies Condition (2). Otherwise, some paper segments for s_j have to go out from the folded state for either terminator. Then, the maximum crease width exceeds k at the crease, which contradicts the assumption. Now, we show Condition (3) by contradiction. We assume that a folder holds some paper segments corresponding to at most two x_js. Then, by Condition (2) and the pigeon hole principle, some folder must have the paper segments corresponding to at least four x_js. Thus we assume that paper segments of four elements, $x_{j_1}, x_{j_2}, x_{j_3}$, and x_{j_4}, are inserted in this folder. Then, since the maximum crease width at this folder is at most k, we have $w(a_{j_1}) + w(a_{j_2}) + w(a_{j_3}) + w(a_{j_4} \leq B$. On the other hand, $w(a_{j_1}) + w(a_{j_2}) + w(a_{j_3}) + w(a_{j_4}) > 4(B/4) = B$ by the condition of the input of 3-PARTITION, which is a contradiction. Therefore, the condition (3) holds.

Once a folded state of the maximum crease width at most k that satisfies the conditions (1) to (3) is obtained by the MV string s, it is easy to construct an answer to the instance of 3-PARTITION from the folded state. □

Since Lemmas 6.2.7 and 6.2.8 are obtained, we finally prove Theorem 6.2.3. By these two lemmas, we found that an instance of 3-PARTITION has a solution if and only if the constructed one-dimensional origami has a folded state of maximum crease width k. Therefore, Theorem 6.2.3 holds.

6.2.3 *Tractability for Bounded k*

In this section, we show an algorithm that decides whether the total crease width is at most k or not in $O((k+1)^k n)$ time. In a technical term, the total crease width problem is said to be *fixed parameter tractable (FPT)* (by this algorithm). To explain in more detail, the total crease width problem can be solved efficiently when the total crease width k is small enough because the algorithm decides it efficiently.

Note that we have to give the value of k explicitly to the algorithm in advance. In other words, we try to execute the algorithm with anticipation that the total crease width is sufficiently small, we can solve the problem if the algorithm succeeds, and otherwise, we find that the instance is intractable. Actually, if k gets bigger, $(k+1)^k$

becomes enormously big. Considering this fact, we can run the algorithm for each $k = 1, 2, 3, \ldots$ in this order, and we can decide it is tractable if the computation halts in a reasonable time, and otherwise, the instance is intractable to be solved.

Although it may seem like a somewhat cryptic algorithm, it can be said that this is the best result, as the computational complexity of this problem is not shown so far. Anyway, let us show the algorithm for finding out if the total crease width is at most k in polynomial time of n for given MV pattern s of length n and the natural number k.

First, let us assume that an MV pattern $s = s_1 s_2 \cdots s_n$ and a total crease width k are given as input. Then, consider the partial pattern $s^{(i)} = s_1 s_2 \cdots s_i$ for each $i = 1, 2, 3, \ldots, n$, and enumerate all folded states of crease width at most k. Specifically, for each folded state up to $s^{(i-1)}$, connect the last ith segment inserted in the possible locations to generate all the folded states for $s^{(i)}$. Representing this idea in the form of a concrete algorithm can be described in Algorithm 5 and the procedure fold(P, i, k). When an MV pattern s and a natural number k are given as input, the algorithm outputs all of the flat folded states of s of total crease width at most k if there exists. If such a folded state does not exist, the algorithm halts without any output.

Input : $s \in \{M, V\}^n$ $(n \geq 1)$ and a natural number k
Output: All flat folded states P consistent with s and of total crease width at most k

```
Initialize P with the 0th segment placed on the interval [0, 1];
fold(P, 1, k);
// Connect the first segment to the current P. Total crease
   width is at most k.
```

Algorithm 5: Algorithm to enumerate all flat folded states of total crease width at most k

```
if i = n then output P else
    take the ith segment;
    foreach Perform the following on the gap between each layer of the current folded state
    P do
        if the ith segment can be inserted into the current gap then
            Let P be the new folded state in which the ith segment is placed in the gap;
            Let k' be the crease width at the ith crease;
            Let j be the number of creases at which the crease width increases by inserting
            the ith intercept;
            // If k − (k' + j) ≥ 0, it is still a feasible flat folded
               state.
            fold(P, i + 1, k − (k' + j));
        end
    end
end
```

Procedure fold(P,i,k)

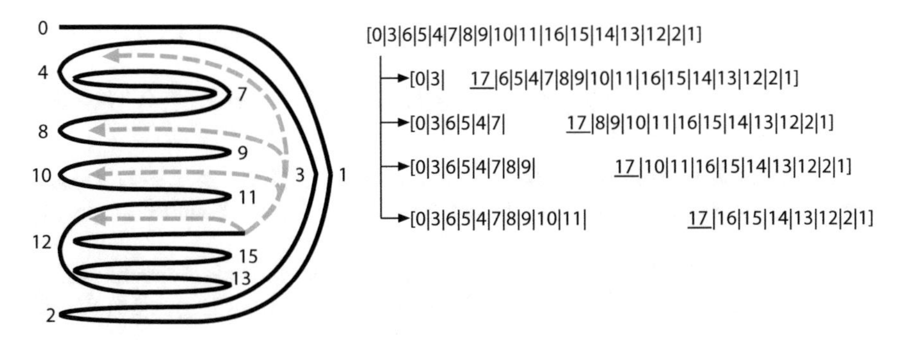

Fig. 6.8 Options that `fold`$(P, 17, k)$ can choose

Example 6.2.9 Let given MV pattern be $MMVVVMMVMVMVVMVMV\cdots$ and the current folded state [0|3|6|5|4|7|8|9|10|11|16|15|14|13|12|2|1] shown in Fig. 6.8. Since the next crease s_{17} is V (assuming that there is still a margin in k), the possible choices by `fold`$(P, 17, k)$ are indicated by the dotted lines in the figure. Depending on each choice, the crease width increases by one at each crease where the paper segment is inserted. On the other hand, the 17th crease increases the total crease width by 0,2,4, and 8, respectively, depending on the chosen location.

This algorithm is a fairly naive algorithm in a sense. It tries every possible way of folding of total crease width less than or equal to k. Therefore, the correctness of the algorithm is almost trivial. When roughly estimating the running time of this natural algorithm, we can immediately obtain a simple upper bound $O(n!)$ because the first segment is uniquely placed, the second segment has two possible places, and the third segment has three possible places, and so on. However, according to Sterling's formula, since

$$n! \sim \sqrt{2\pi n}\left(\frac{n}{e}\right)^n,$$

it actually gives an upper bound $O(n^n)$, which is exponential time. On the contrary, if we evaluate the running time more carefully, we can obtain the upper bound $O((k+1)^k n)$ of the running time, as indicated in Theorem 6.2.4. Let us consider the running time of subroutine `fold`(P, i, k) more carefully.

First, consider the case where the variable k' and j in the algorithm satisfy $k' + j = 0$ by inserting the ith segment. This happens when the $(i-1)$st segment is at the top or bottom, visible from the outside of P, and the crease width made by the ith segment is 0. In other words, this is when the ith segment is folded outside of the $(i-1)$st segment which is visible from the outside of P (or when $i = 0$).

Except in the case of above, $k' + j > 0$ holds wherever the ith segment is inserted. Let h be the number of places in the folded paper layers where ith segment can be inserted, and $L_1, \ldots, L_h$ be the places from the one closer to the $(i-1)$st segment. Then we have the following:

- When we insert the ith segment to L_ℓ for $\ell \in \{1, \ldots, h-1\}$, we have $k' \geq \ell - 1$ and $j \geq 1$, or $k' + j \geq \ell$.
- When we insert the ith segment to L_h, we have $k' \geq h - 1$ with $j \geq 0$. Therefore, we have $k' + j \geq h - 1$.

By this discussion, we have $k - (k' + j) \geq 0$ only if $i \in \{1, \ldots, k, k+1\}$. In other words, the number of branches in the call to `fold`(P, i, k) is bounded at most by $k + 1$.

To summarize the above cases, the following recurrence formula is obtained for the worst-case time complexity $t(n, i, k)$ of the number of times of calling `fold`(P, i, k):

$$t(n, i, k) \leq \begin{cases} O(1) & i = n \text{ or } k < 0. \\ (k+1)(t(n, i+1, k-1) + O(n)) & \text{Otherwise.} \end{cases}$$

Solving this recurrence formula yields $t(n, i, k) = O((k+1)^k n)$. Therefore, Theorem 6.2.4 holds.

References

[Gar67] M. Gardner, Mathematical games. Sci. Am. **216**(3), 124–125 (1967); **216**(4), 118–120 (1967); **217**(1), 115 (1967)

[Gar08] M. Gardner, *Origami, Eleusis, and the Soma Cube: Martin Gardner's Mathematical Diversions* (Cambridge University Press, The New Martin Gardner Mathematical Library, 2008)

[GJ79] M.R. Garey, D.S. Johnson, Computers and Intractability – A Guide to the Theory of NP-Completeness. Freeman (1979)

[LZ77] J. Ziv, A. Lempel, A universal algorithm for sequential data compression. IEEE Trans. Inf. Theory **23**(3), 337–343 (1977)

[LZ78] J. Ziv, A. Lempel, Compression of individual sequences via variable-rate coding. IEEE Trans. Inf. Theory **24**(5), 530–536 (1978)

Part IV
Advanced Problems

In Part IV, we introduce four topics of the computational origami. They are on the front-line of this research area. Comparing with Part II and Part III, they are somewhat limited or developing, rather meaningful to provide challenges that are expected to develop in the future. It is at the forefront of "just recently known", so the difficulty is unexpectedly disjointed. Some topics can be research themes at graduate schools; on the other hand, there are some themes that high school students can be likely to solve if good ideas emerge. It is a challenge to readers, in other words, I want you to have fun, troubles, or consider depending on the level of you.

The first two topics are related to the nets of polyhedra.

The first one is dealing with polyhedra such as triangular pyramids and quadrangular pyramids, that is, general pyramids. The research of this problem was initiated by myself when I was asked about the folding problem of triangular pyramids and quadrangular pyramids by one of my puzzle friends, Teruo Nishiyama, who is a former mathematics teacher. The nets of a triangular pyramid are relatively easy, but the problem becomes very difficult for a quadrangular pyramid or pyramids with more vertices. It is a simple and natural problem, and it seems that there can be existent some results; however, as far as I investigated, it does not seem to be well studied so far. In the chapter, we give the features of the nets of these general pyramids, consider the "bumpy" case where the bottom surface is not flat, and give some algorithmic solutions.

In the second topic, we investigate special nets which are unfolded by a method called "zipper unfolding". It may be imagined from the name, but in short, a zipper unfolded net is a net obtained by cutting along a line that you can draw in one stroke. That is, once you insert scissors in the polyhedron, you cut it along a line without branching off from it, and unfold it as it is. Since it can be realized by attaching only one zipper, it is called by such a name. In the chapter, let us give concrete examples of polyhedra that cannot be zipper unfolded, and introduce interesting open problems. This area may seem easy, but there are many problems that have not yet been solved.

The other two topics are related to computational complexity of the folding algorithm.

The third topic is a generalization of the crease width minimization problem of stamp folding of Sect. 6.2. In this generalization, the creases are set to general intervals instead of equal intervals. Originally, "crease width" is defined by the number of sheets of paper layers at a crease; however, in general intervals, creases do not necessarily overlap with the same place anymore. Thus, it is necessary to generalize from the definition itself, and the property of the problem has changed a lot.

The last topic is an undecidable folding problem. This is a study of computability of a computation model, and it is a rather theoretical result that is possible to create an undecidable problem on origami when we introduce some computation model. It is recommended for readers who like arguments close to logic such as diagonalization and incompleteness theorem or pure mathematics. You may feel the depth of the notion of origami.

Chapter 7
Bumpy Pyramids Folded from Petal Polygons

Abstract In this chapter, we consider a special set of polygons and convex polyhedra folded from it. After giving the (counterintuitive) answers to the puzzle given in this book, we consider the folding problem of (bumpy) pyramids folded from a special set of polygons called "petal polygons".

7.1 Convex Polyhedra Folded from a Polygon

First, let me point out that there are many convex polyhedra that can be folded from a polygon. For example, as also pointed in Chap. 1, you can fold 23 different convex polyhedra from one polygon called Latin cross.

Also, as described in Chap. 1, even for a cube, there are many nets that consist of six squares and are not listed in Fig. 1.1. I first give the answers to this interesting puzzle here.

Difficulty of nets of a convex polyhedron (Answer)

First, we consider the nets of a cube that consist of six squares, but their sizes are not necessarily the same. One example is shown in Fig. 7.1a. If the size of the cube is $1 \times 1 \times 1$ and the surface area is 6, this net consists of two squares of area 2 and four squares of area 1/2. Puzzle researcher Naoaki Takashima confirmed that there are 54 nets in total based on this idea.[1] Second, we consider the nets that consist of six congruent squares. In this case, once you notice that "creases and cut lines do not necessarily have to follow the edges of the cube", you can come up with the solutions as in Fig. 7.1b.[2] Since this shift width is an arbitrary real number, the solution exists infinitely.

R. Uehara, *Introduction to Computational Origami*,
https://doi.org/10.1007/978-981-15-4470-5_7

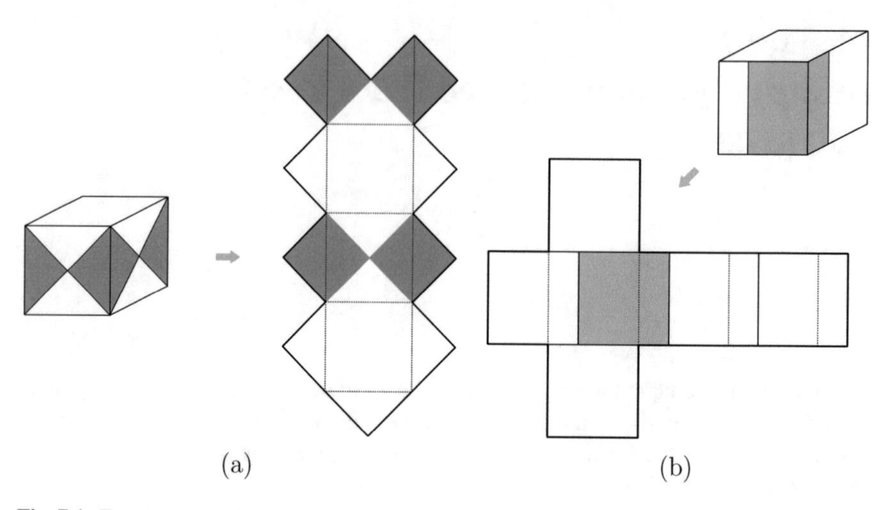

Fig. 7.1 Two unexpected nets of a cube

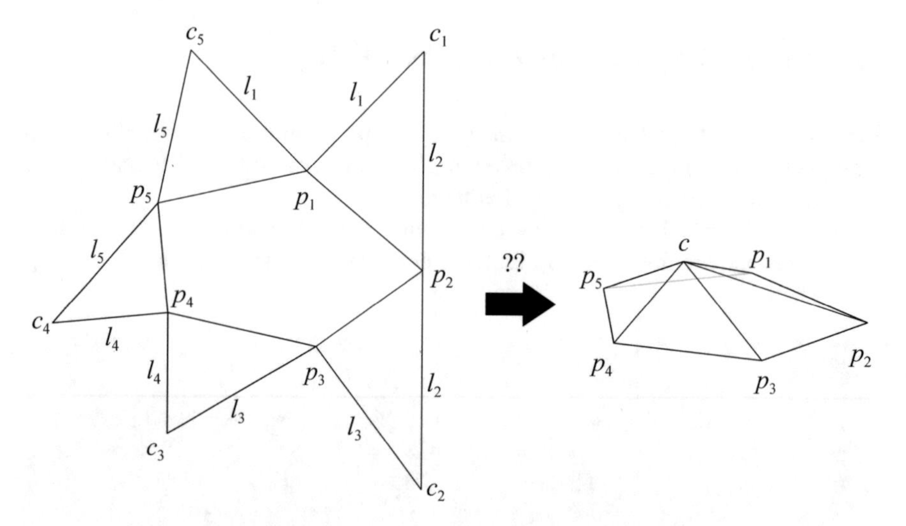

Fig. 7.2 Pyramid obtained by folding a petal polygon

7.2 Petal Folding Problem

Now, we turn to the main topic. In this section, we consider the problem of a pyramid folded from a star-shaped polygon, or conversely, a polygon which can be obtained by unfolding a pyramid from its apex. You may want to call these polygons "star

[1]It is published in his blog at http://www.iwa-masaka.jp/56290.html of Masaka Iwai who is the inventor of this puzzle.

[2]This answer is also listed in Masaka Iwai's blog http://www.iwa-masaka.jp/56291.html.

polygons", but since this term is sometimes used in other contexts, we will call them "petal polygons" in this book (Fig. 8.3). Let us define it exactly. When a polygon $P = (p_1, c_1, p_2, c_2, \ldots, p_n, c_n)$ satisfies the following three conditions, it is called *petal polygon*:

1. The polygon $B = (p_1, p_2, \ldots, p_n)$ is convex, and each point c_i is outside of B. We call B as *base* of the petal polygon.
2. The two edges connected to each point p_i have the same length. That is, $|p_i c_i| = |c_{i-1} p_i| = \ell_i$ holds for each $i = 1, 2, \ldots, n$.[3]
3. The sum of interior angles in c_i is less than 360°. Namely, $\sum_{i=0}^{n-1} \angle p_i c_i p_{i+1} \leq 360°$ holds.

Using the above term, we introduce *petal folding problem* discussed in this section. We are given a petal polygon. For each vertex p_i of this polygon, there are two edges of the same length connected to it, so we glue each pair. Then we have a polyhedron, say, P. Of course, we cannot say whether P always exists so far. If it exists, as a result of gluing, we call the vertex which appears at the top *apex vertex* c. In this section, we will consider three problems for this petal folding problem.

We first examine the case that the base B is flat after folding. That is, we consider the conditions of a petal polygon that can be folded into a pyramid with flat base B, or without any folding of the base B. Let us call this problem the *petal pyramid folding problem*. That is, for a given petal polygon P, the problem asks to decide if P can be folded into a pyramid with the flat base B. To show the conclusion earlier, this problem can easily be solved when you notice a certain fact.

Theorem 7.2.1 *For a given petal polygon P, the petal pyramid problem can be solved in linear time.*

Now let us consider the case that P cannot be folded into a pyramid as a result. Intuitively, even if P is not folded into a pyramid, when you forcefully glue the corresponding edges and gather all the points c_i at the apex c, it seems that some kind of polyhedron may be folded. We next consider these problems. Specifically, if you put crease lines in B properly, P is likely to be folded into a polyhedron. As a division of B, we consider *triangulations* of B. (The details of the notion of a triangulation will be discussed shortly.) Suppose you fold all edges $p_i p_{i+1}$ in valley and collect c on your side against B. Then, if you properly choose the triangulation of B and assign mountain folds and valley folds to it, a closed polyhedron can be folded. We call this polyhedron *bumpy pyramid*. The bumpy pyramids are not obvious at all, but they have very interesting properties. We will learn them in this section. For example, even when $n = 4$, that is, B is a quadrangle, it shows interesting behavior. Typically, we can obtain two different bumpy pyramids folded from P; one is a convex polyhedron, and the other is a concave polyhedron. However, only convex one can be folded in some cases, and no polyhedron can be folded in some cases.

As we will see later, in general n, the number of ways of folding increases exponentially. Therefore, as the next problem, we will consider the following two problems.

[3]We take modulo n. That is, $n + 1$ is assumed to be 1.

We first consider whether or not any convex polyhedra can be folded from a given petal polygon. For this problem, we have an interesting result.

Theorem 7.2.2 *For a given petal polygon P of $2n$ vertices, whenever any polyhedron can be folded from P, a convex one can be folded. Moreover, the crease lines for it can be computed in linear time.*

Although there exists an exponential number of polyhedra, the results are very interesting that a convex one is always present among them, and that it can be obtained in linear time. We will use properties related to generalized Voronoi diagrams called power diagrams to prove Theorem 7.2.2. More precisely, we will show that the bumpy pyramid folding corresponds to the power diagram in a non-trivial way. We will also describe this notion later.

Here, I will point out the relationship between this problem and Alexandrov's theorem. Roughly, the Alexandrov's theorem claims the following.

Theorem 7.2.3 (Alexandrov's Theorem) *Given a geometric structure of a convex polyhedron, its shape is uniquely determined.*

The detailed definitions go beyond the scope of this book, so I would like you to refer to [DO07], for example. In the context of the bumpy pyramid problem, the theorem claims that, if a convex polyhedron can be folded from a given P, its shape is uniquely determined. The theorem itself was proved in 1942, but the constitutive proof was given by Bobenko and Izmestiev [BI08], and a pseudopolynomial time algorithm was given by Kane et al. [KPD09]. That is, Theorem 7.2.2 can be computed by the pseudopolynomial time algorithm by Kane et al. However, the upper bound of the running time of this algorithm is $O(n^{456.5}r^{1891}/\epsilon^{121})$ time (where r is the ratio between the farthest distance and the nearest distance of the vertices, and ϵ is the relative accuracy of coordinates), which is quite a theoretical polynomial time algorithm and far from practical one. As far as I know, Theorems 7.2.1 and 7.2.2 are the first efficient algorithms of Alexandrov's theorem for non-trivial convex polyhedra.

Next we consider the problem of folding a polyhedron of *the largest volume* among bumpy pyramids. At first sight, if there exists a convex bumpy pyramid, it seems that it has the largest volume as well. Surprisingly, however, that intuition is wrong in general. We here give two simple counterexamples in Fig. 7.3. You may calculate their volumes seriously, but for these two examples, we can give a bit more intuitive and simple explanations. We first consider the example in Fig. 7.31. To fold it in a convex way, you make a mountain fold along the line p_1p_3. Since the resulting bumpy pyramid is very thin, the volume is almost zero. On the other hand, if you make a valley fold along the line p_2p_4 and dent the base inward, the bumpy pyramid can realize a much larger volume. In the other figure of Fig. 7.32, when you make a mountain fold along the line p_2p_4, it is almost the same as the triangular pyramid having the base triangle $p_1p_2p_4$. In this case, the point p_3 can be regarded as an inside point of the triangle p_2cp_4. Alternatively, when you make a valley fold along the line p_1p_3, this dent is not so large that the polyhedron becomes a slightly concave quadrangular pyramid having the base square. Thus, this concave

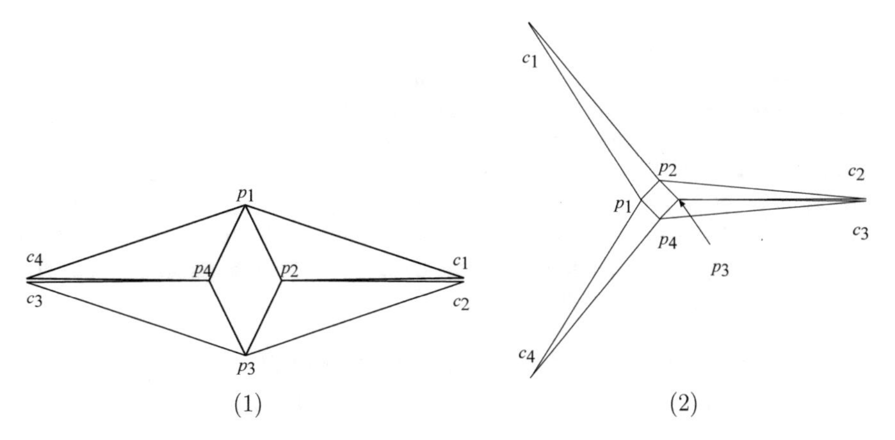

Fig. 7.3 Two counterintuitive examples of the bumpy pyramid folding: concave bumpy pyramids have larger volumes than convex ones

quadrangular pyramid has the double volume of the convex triangular pyramid. Both examples are easy to understand by folding by yourself.

As will be shown later, the formula for computing the volume of a tetrahedron is known when all lengths of the edges are given [Sab98]. Therefore, when a triangulation of the base B is given, the volume of the bumpy pyramid can be computed by the sum of all volumes of the tetrahedra defined by each triangle of the triangulation. In general, however, it is not a good idea to compute the maximum volume by computing the volumes for all cases since the number of triangulations of the base B is exponential. In this section, we show an efficient algorithm using dynamic programming.

Theorem 7.2.4 *When a petal polygon P with $2n$ vertices is given, the maximum volume of the bumpy pyramid folded from P can be computed in $O(n^3)$ time.*

Intuitively, it seems that the convex bumpy pyramid has the maximum volume in many cases. That is, the examples in Fig. 7.3 are considered to be exceptional cases in a sense. However, the characterization is not known.

Open Problem 7.2.1 *In what case does the convex bumpy pyramid folded from a petal polygon P have the largest volume? Or what is the condition that the convex bumpy pyramid does not have the maximum volume?*

7.3 Triangulation, Voronoi Diagram, and Power Diagram

We first introduce some necessary notions of computational geometry in this section. (In this book, knowledge on computational geometry is not prerequisite, so we briefly introduce only the necessary notions. Readers interested in further details, see [BCK+10] for example).

For a given polygon B, draw straight lines inside of this polygon until it cannot be drawn any more. At this time, each straight line segment connects the two vertices of B, and it should not cross the edges of B or the straight lines already drawn. Since you draw lines until you cannot draw any longer, all inner faces of B become a triangle. (Otherwise, you can still draw a line.) Then, this set of straight line segments is a *triangulation* of B. Hereafter, we only consider the case where B is convex. Let T be any triangulation of B. For this T, we consider the graph $G(T) = (V, E)$ which expresses the adjacency relationship of the triangles in this triangulation. Precisely, we define the graph $G(T)$ as follows:

$$V = \{v \mid v \text{ is one of the triangles in the triangulation of } B\},$$
$$E = \{\{u, v\} \mid u \text{ and } v \text{ share a line segment of the triangulation}\}.$$

Intuitively, each triangle corresponds to a vertex, and if two triangles are adjacent to each other, they are connected by an edge. We call $G(T)$ defined in this way as *dual* of T.[4] Since the dual is a relationship used in both directions, conversely, T is sometimes called dual of $G(T)$. Then we have the following.

Theorem 7.3.1 *Let n be any natural number greater than 3, and consider a triangulation T of a convex polygon $B = (p_1, p_2, \ldots, p_n)$ of n vertices. Then, for the dual graph $G(T)$, (1) $G(T)$ is a tree, and (2) $G(T)$ has $n - 2$ vertices.*

Since this theorem is a basic theorem of computational geometry, we show a brief proof. For interested readers, see the literature [BCK+10], for example.

Proof (*Sketch*)
(1) It is enough to show that $G(T)$ is connected and acyclic. $G(T)$ is almost trivially connected since B is a convex polygon. We can show that $G(T)$ is acyclic by contradiction. If $G(T)$ has a cycle, triangles of triangulation of B are connected in cyclic. In that case, the outer and inner regions are divided by the cycle. Then, we have a vertex in the area inside of the original B, which contradicts with the assumption that B is convex. Therefore, T is acyclic and it turns out to be a tree.

(2) We use an induction for the number n of vertices of B. Since the case $n = 3$ is trivial, we assume $n > 3$. Then, by (1), $G(T)$ is a tree. Any tree has at least two leaves. (This is a basic fact in graph theory. Readers who need proof, for example, see [Die96].) The triangle t corresponding to this leaf ℓ is adjacent to only one triangle on B. This means that, out of the three edges of t, the two edges are also edges of B. That is, t is a triangle consisting of three consecutive vertices on B. Let p_{i-1}, p_i, p_{i+1} be these vertices. The polygon $B' = (p_1, p_2, \ldots, p_{i-1}, p_{i+1}, \ldots, p_n)$ obtained by removing p_i from B has a triangulation corresponding to the tree T' which is obtained from T by removing the leaf ℓ. Therefore, the theorem is established by induction. □

In the proof of Theorem 7.3.1, we show that any triangulation T of B has at least two triangles t using three consecutive vertices of B if $n > 3$. We call such a triangle

[4] Also see the dual of a polyhedron appearing in Sect. 2.1.

Fig. 7.4 An example of a Voronoi diagram

an *ear* of the triangulation T. That is, any triangulation of a convex polygon with at least four vertices has at least two ears.

Next, we introduce *Voronoi diagram*. We construct the Voronoi diagram for a given set of n points $P = \{p_1, p_2, \ldots, p_n\}$. Each element p_i is referred to as *site* of the Voronoi diagram. The Voronoi diagram is obtained by dividing the plane from the viewpoint of which site is the closest. To put it more precisely, the Voronoi diagram $V(P)$ for a given point set P is a partition of the region that satisfies the following conditions: (Fig. 7.4).[5]

- Each region contains exactly one site.
- At each point in a region, the closest site is the site contained in the same region.

A point equidistant from two or more sites is on a boundary line in the Voronoi diagram. This boundary line is a straight line segment or a half straight line, and it is called *Voronoi edge*. It is not difficult to see that a Voronoi edge is the perpendicular bisector of the closest sites on both sides. For example, when three Voronoi edges are connected to a single point, you can see that this point is equidistant from the three sites. Such a point is called *Voronoi node*.

[5] Voronoi diagram plays a very important role in computational geometry and has many applications. Therefore, international conferences on Voronoi diagram as the main theme are held regularly. If you search on the web, you can find many scripts that generate Voronoi diagrams. Figure 7.4 is generated by a script of Alex Beutel's webpage (http://alexbeutel.com/webgl/voronoi.html).

In this book, we use *power diagram* which generalized this Voronoi diagram. (In this book, only the necessary properties are considered, but for further details, see, e.g., [Aur87].) In the Voronoi diagram, a certain Voronoi edge is a perpendicular bisector of two sites. In the context of the Voronoi diagram, it can be said that it is equidistant from both sites. Here, consider giving "weight" to the sites. For example, suppose that points p_i and p_j have weights $w(p_i) = x$ and $w(p_j) = y$, respectively. In the Voronoi diagram, all of the sites have the same weight, for example, $w(p) = 1$ for all sites p. In the power diagram, the Voronoi diagram is extended so that the distance to the line ℓ between p_i and p_j is proportional to their weights. That is, in this case, any point q on ℓ satisfies $|p_i q| : |p_j q| = w(p_i) : w(p_j)$. This can be said to be a natural extension when considering applications such as facility location.

In this section, we need to compute the power diagram for a point set in convex position. For this problem, a linear time algorithm is given by Aggarwal et al. [AGS+89]. For a given set of sites in convex position, we consider a Voronoi edge of the power diagram. Then we have the following lemma.

Lemma 7.3.2 *For the power diagram of a given set of sites in convex position, we consider a graph $G = (V, E)$, where V is the set of Voronoi nodes and E is the set of Voronoi edges. Then G is connected and acyclic. Moreover, each vertex in G has degree 3.*

Proof Since the arrangement of sites is convex, there are no sites inside. Therefore, the Voronoi edges do not form a cycle inside of the convex arrangement. Also, even outside, the edges do not intersect each other, so the power diagram for the set of sites in convex position is a connected graph without cycle. Since the position of each site is general position, the degree of each Voronoi node is 3. □

Roughly speaking, if you consider a Voronoi edge as an edge and a Voronoi node as a vertex, the graph composed of them is a "tree". However, since the line passing between the two adjacent sites on the original convex position is a perpendicular to the line segment connecting these two sites, the perpendicular continues to infinity. Therefore, this "tree" has no so-called leaves. In other words, it can be considered as a tree such that all the leaves are at infinity.

7.4 Preparation of Petal Pyramid Folding

When we introduced the notion of a petal polygon in Sect. 7.2, we listed three conditions. The third condition is that "the total angle at the apex c is less than or equal to 360°" has an important meaning when $n > 3$, so we consider it here. If the total of the interior angles c_i exceeds 360°, the curvature at this apex is a negative value, or the apex becomes a *saddle point* in any polyhedron you fold. A simple example is shown in Fig. 7.5. When you fold the petals around B and along the diagonal $p_1 p_3$, the apex c becomes the saddle point whose surrounding angle exceeds 360°. Especially in the first two problems, we only consider convex polyhedra, so it can be excluded

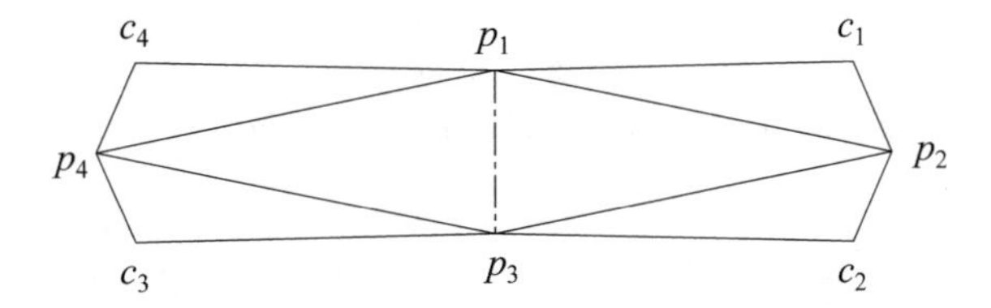

Fig. 7.5 This net does not satisfy Condition 3, and the vertex c becomes a saddle point

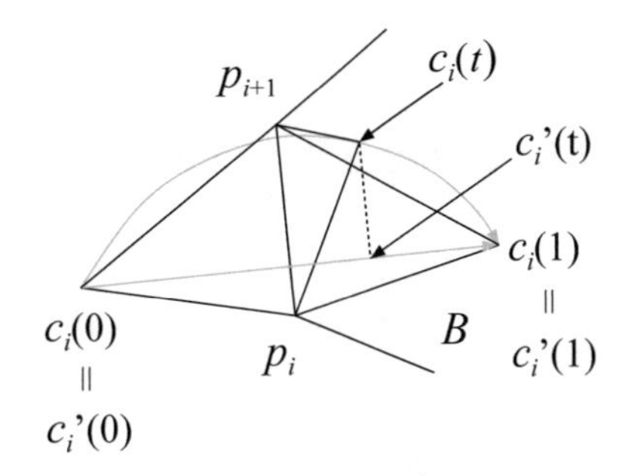

Fig. 7.6 Trace of $c_i'(t)$ when $\triangle p_i p_{i+1} c_i(t)$ is folded

from the beginning when the apex becomes a saddle point. In a sense, this can be considered a case where the petals are too short. In the last problem (maximization of volume), although it may be meaningful to consider such a situation, we leave this as an open problem.

Next, we consider the folding process in more detail. First, we assume that P is placed on the xy-plane, and each point p of P is given by the coordinate $(x(p), y(p))$. In the general folding state, P is in the three-dimensional space, so each point p is represented by the coordinates $(x(p), y(p), z(p))$. On a plane or space, for given two points p and q, we denote by $\overline{pq}$ the line without an endpoint including a line segment pq with both endpoints p and q. Let us denote a triangle consisting of three points $p_i p_{i+1} c_i$ of P by T_i. We consider folding this triangle T_i along the line $\overline{p_i p_{i+1}}$ onto the base B. When T_i is folded to the opposite side completely and it overlaps with B, T_i is flipped along the line $\overline{p_i p_{i+1}}$; however, we consider the intermediate position as a function of a time. That is, T_i is not folded at all at time $t = 0$, T_i is flipped along $\overline{p_i p_{i+1}}$, and it overlaps B at time $t = 1$ (Fig. 7.6). In this flipping of the triangle T_i, we consider that the point c_i moves from $c_i(0)$ to $c_i(1)$. If this c_i reaches the apex c on the way and it is glued to form a (bumpy) pyramid, we have $c = c_i(t)$ at a certain time t with $0 < t < 1$. Let c_i' be the point projected on the xy-plane when c_i is moving. That is, this projection is $c_i' = (x(c_i), y(c_i), 0)$ when $c_i = (x(c_i), y(c_i), z(c_i))$. Now we consider the *trace* τ_i of c_i' which is drawn by c_i' on xy-plane when c_i moves from $c_i(0)$ to $c_i(1)$. Imaging the actual movement, you can see that τ_i is a straight line segment $c_i(0)c_i(1)$ on the xy-plane. In the following, rather than the trace from $c_i(0)$ to $c_i(1)$ drawn on space when the point c_i moves on space, we consider the trace τ_i drawn by projecting it on the xy-plane.

Lemma 7.4.1 *For the base polygon B, we fold the triangle T_i along the line $\overline{p_i p_{i+1}}$ until T_i overlaps B. Then the trace τ_i drawn on the plane by $c_i'(t)$ is the chord joining* *two intersection points of the circle C_i of the radius $|p_i c_i|$ centered on the point p_i and the circle C_{i+1} of the radius $|p_{i+1} c_i|$ centered on the point p_{i+1}. In other words, τ_i is the line segment on the perpendicular of the line $\overline{p_i p_{i+1}}$.*

Now, this lemma can be easily seen as if proof is unnecessary. However, the property shown in this lemma is very useful and important. We also note that this trace τ_i can be easily constructed and computed by the lemma.

7.5 Folding Pyramid

In this section, we will consider the case where the base B is fixed on the xy-plane and each point c_i is (valley) folded so that the z-coordinate becomes positive. Then, the trace projected on the xy-plane is τ_i. (Intuitively, we collect the points c_i to make an apex c on our side against B. Then $p_i = (x(p_i), y(p_i), 0)$ in the initial flat state, and $c_i = (x(c_i), y(c_i), z(c_i))$ always satisfies $z(c_i) \geq 0$.)

By Lemma 7.4.1, clearly, in order for all the points c_i to reach the common apex c, all the corresponding traces τ_i meet at c' on the plane. However, this is redundant; we can prove that if the traces of $n-1$ points meet at a common point, the last nth point always reaches there and intersects.

Lemma 7.5.1 *Suppose that the base B of the petal polygon P is an n-gon. Then, the necessary and sufficient condition that P can fold into the pyramid (of the flat base B) is that $n-1$ traces τ_i meet at a common point c'.*

Proof Let us consider a three-dimensional space and assume that each point p_i is on the xy-plane. Here, considering the sphere S_i of the radius ℓ_i centered on each point p_i, S_i goes through the points c_{i-1} and c_i. (From the other viewpoint, c_{i-1} and c_i are the points on the sphere S_i.) We can consider that each trace τ_i is a projection of the disk which is the intersection $S_i \cap S_{i+1}$ of the two spheres S_i and S_{i+1} onto the xy-plane. Assuming here that $n-1$ traces intersect at a common point c', then all of the spheres pass through the point c', and conversely the point c' is on all the spheres. Therefore, the last remaining trace also passes through the point c' as the corresponding two spheres pass through this point. Thus, if $n-1$ traces intersect at one point c', we can see that all n traces intersect at that point c'.

First, if P can fold into a pyramid, all c_i should gather at common apex c. That is, for each $i = 1, \ldots, n$, there exists $0 < t_i < 1$ such that $c_i(t_i) = c$. Therefore, by Lemma 7.4.1, the projection c' of c is on all the traces τ_i.

Next, we show the opposite. For each i, if the point c' passes the trace τ_i, there exists $0 < t_i < 1$ such that $c_i'(t_i) = c'$. Here, since the point $c_i(t_i)$ is a point on the intersection of the two spheres $S_i \cap S_{i+1}$, these two spheres are located at the same height above c'. Similarly, the point $c_{i-1}(t_{i-1})$ is the point on the intersection of two spheres $S_{i-1} \cap S_i$; thus, these two spheres go through the same height above c'. By the transitive law, all spheres pass through a point at the same height above c', and every $c_i(t_i)$ goes through the same point c, so this is the apex of the folded pyramid. □

To prove Theorem 7.2.1, it is sufficient to show that we can find the common point c' shared by all traces in linear time. The candidate for this point is the solution of the system consisting of n linear equations $(c' - c_i) \cdot (p_i - p_{i+1}) = 0$. By Lemma 7.5.1, we can examine $n - 1$ equations among them, for example, when $n = 3$, the candidate point is unique at any time. In the case of $n > 3$, you can select any two linear equations and check whether its solution satisfies other equations. Once you find the candidate point c' in this way, you can compute the distance from c' to each line segment $\overline{p_i p_{i+1}}$, and confirm that it is shorter than the distance from c_i to $\overline{p_i p_{i+1}}$. This can be done in the following way:

$$\left|(p_i - p_{i+1})^{\perp} \cdot (c' - p_i)\right| \leq (p_i - p_{i+1})^{\perp} \cdot (c_i - p_i).$$

The above computation can be done in linear time as a whole, so we can prove Theorem 7.2.1.

Based on Theorem 7.2.1, you can draw a net of a pyramid inversely. That is, first, place the convex polygon B of the base on the xy-plane and set the projection of the point you want to be the apex inside of B as the origin o. Next, decide one triangle flap so that c_i is sufficiently long (or the desired length). Then you can obtain the pyramid where the apex c comes above o.

Theorem 7.5.2 *Let $B = (p_1, p_2, \ldots p_n)$ be a convex polygon on the xy-plane, and o be the origin inside. For any i, decide a certain length ℓ_i that satisfies $\ell_i > |p_i o|$. Then there is a petal polygon with $|p_i c| = \ell_i$ that can be folded into a pyramid in which the projection of the apex c becomes the origin o. Also, we can compute the sequence $(p_1, c_1, p_2, c_2, \ldots p_n, c_n)$ of vertices in linear time.*

Proof Suppose we decided to make the length ℓ_1 to be $\ell_1 > |p_1 o|$ without loss of generality. First consider the triangle op_1c. Since we set $\ell_1 > |p_1 o|$, we have $\ell_1 = |p_1 c|$, and since we know the length $|p_1 o|$, the height of the apex c, which is $|co|$, can be computed by $\sqrt{\ell_i^2 - |p_i o|^2}$. By using this height $|co|$ of c as a reference point, we can compute $|cp_i|$ for each i by $|cp_i| = \sqrt{|co|^2 + |op_i|^2}$ supported by the triangle op_ic. Obviously, since $|op_i| < |cp_i| = \ell_i$, we can construct a petal polygon from ℓ_i computed in this way. □

7.6 Four-Vertex Bumpy Pyramid Folding

Hereafter, we consider folding bumpy pyramids, that is, pyramids obtained by folding along the diagonal lines connecting the two vertices of the base. To simplify, we assume that the base of the bumpy pyramid is completely triangulated. In other words, it is assumed that no quadrangle is remaining flat on the base, and the base is completely triangulated, and every diagonal is folded in mountain or valley.

First, when $n = 3$, the base of the petal polygon is a triangle, and it can be considered that the length of the tetrahedron's six edges (and its geometric relationship) is given. If a tetrahedron exists for the six edges of given lengths, its shape is uniquely determined by the Alexandrov's theorem. Then, we can compute the volume in constant time if we use the formula in [Sab98]. In fact, as you can see from the column below, when $n = 3$, you can use the formula instead of Theorem 7.2.1. You can compute the volume from the length of each edge of a given petal polygon P and determine whether it has a real positive solution.

Formula of volume of a tetrahedron

In [Sab98], Sabitov showed equations that give volumes of more general polyhedra; however, the equation for the volume of a special three-dimensional shape tetrahedron has been known for a long time. Here we show a concrete formula. Let c be the apex, and p_1, p_2, p_3 be the vertices of the base of the tetrahedron. We denote each length of the six edges by $|cp_1| = \ell_1$, $|cp_2| = \ell_2$, $|cp_3| = \ell_3$, $|p_1p_2| = \ell_4$, $|p_2p_3| = \ell_5$, and $|p_3p_1| = \ell_6$. Then the following equation holds for the volume V of the tetrahedron:

$$
\begin{aligned}
V^2 = \frac{1}{144}(&\ell_1^2\ell_5^2(\ell_2^2 + \ell_3^2 + \ell_4^2 + \ell_6^2 - \ell_1^2 - \ell_5^2) + \ell_2^2\ell_6^2(\ell_1^2 + \ell_3^2 + \ell_4^2 + \ell_5^2 - \ell_2^2 - \ell_6^2) \\
&+\ell_3^2\ell_4^2(\ell_1^2 + \ell_2^2 + \ell_5^2 + \ell_6^2 - \ell_3^2 - \ell_4^2) - \ell_1^2\ell_2^2\ell_4^2 - \ell_2^2\ell_3^2\ell_5^2 - \ell_1^2\ell_3^2\ell_6^2 - \ell_4^2\ell_5^2\ell_6^2).
\end{aligned}
$$

When the value on the right side is positive, V has a real solution, which is the volume. On the other hand, when the value on the right side is negative, there is no feasible volume, that is, the tetrahedron cannot exist for those six edge lengths.

Next, we consider a special case of $n = 4$. The problem has already become non-obvious then, and considering this case is very convenient when we solve the general case.

Let $P = (p_1, c_1, p_2, c_2, p_3, c_3, p_4, c_4)$ be a petal polygon. Then, if you attempt to fold a bumpy pyramid from P, you can fold along one of two possible diagonal lines; folding p_1p_3 and folding p_2p_4. The following theorem holds for these two choices.

Theorem 7.6.1 *For a petal polygon $P = (p_1, c_1, p_2, c_2, p_3, c_3, p_4, c_4)$, one of the following three holds: (1) No bumpy pyramid can be folded. (2) Only one convex pyramid can be folded. (3) One convex pyramid and one concave pyramid can be folded. Furthermore, we can determine which case corresponds to a given P in a constant time.*

Proof First, assume that we fold along the diagonal p_2p_4 of the base and obtain a bumpy pyramid anyhow. Then we can regard this pyramid as the polyhedron obtained

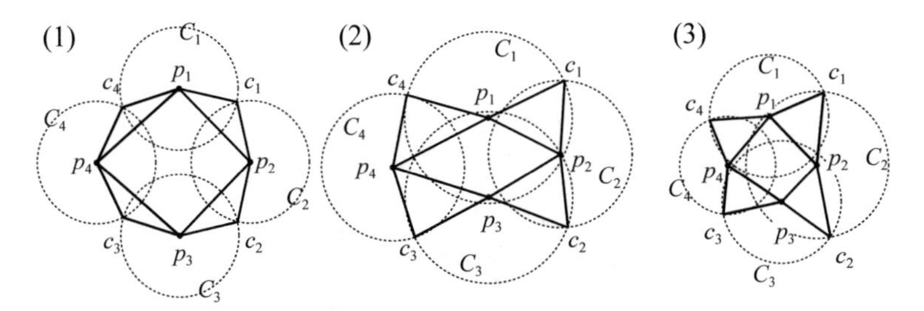

Fig. 7.7 Three possible cases of bumpy pyramids when $n = 4$: (1) No pyramid can be folded. (2) Only one convex pyramid can be folded. (3) One convex pyramid and one concave pyramid can be folded

by gluing two tetrahedra $cp_1p_2p_4$ and $cp_3p_2p_4$ at the common triangular face cp_2p_4. The necessary and sufficient condition that this bumpy pyramid can be folded is that this common triangle forms a triangle properly, that is, three edges $|p_2p_4|$, $|cp_2| = \ell_2$, and $|cp_4| = \ell_4$ satisfy the triangular inequality. Here, let C_i be the circle of radius ℓ_i with the center at p_i. Then the necessary and sufficient condition that the triangular inequality is satisfied is that C_2 and C_4 intersect. The same argument can be made for the circles C_1 and C_3. By the arguments, it turns out that there are the following three cases.

Case 1: Both pairs of circles (C_1, C_3) and (C_2, C_4) do not intersect (Fig. 7.7(1)). This is the case where the heights of the four triangles are short compared with the base. Therefore, you cannot fold into any pyramid.

Case 2: In the case where only one of the two pairs (C_1, C_3) and (C_2, C_4) intersects (Fig. 7.7(2)). We assume that C_1 and C_3 intersect as in Fig. 7.7(2) without loss of generality. In this case, the three points cp_2p_4 are not triangle because the three edges do not satisfy the triangular inequality. Therefore, even if you fold the base along the diagonal line p_2p_4, you cannot make a pyramid. Thus, you can fold into a pyramid only when you fold along the diagonal p_1p_3.

Here, we show that this pyramid is convex. We first suppose that we fold along p_1p_4 and p_3p_4, and glue c_3 and c_4 to c' without folding the diagonal p_1p_3. Then we fold along p_1p_2 and p_3p_2, and glue c_1 and c_2 to c'' (Fig. 7.8). Now it is easy to see that we need to valley fold along p_1p_3 to glue c' and c'' to make the apex c; otherwise, c' and c'' move away.

Case 3: In the case where both pairs (C_1, C_3) and (C_2, C_4) intersect as in Fig. 7.7(3). By Alexandrov's theorem, it is impossible for both to become convex pyramids. That is, both of them are concave, or either one is convex and the other is concave. We prove that the latter one holds.

First, consider the case of folding along the diagonal p_1p_3 (Fig. 7.9). The resulting polyhedron can be regarded as consisting of two tetrahedrons $cp_1p_2p_3$ and $cp_1p_4p_3$ sharing the triangle $T = (p_1, p_3, c)$. This triangle T can be found on the original petal polygon appearing at the intersection of the circles C_1 and C_3 and the line p_1p_3 (thick lines in Fig. 7.9). Then, since there are two intersections between the circles C_1 and C_3, let t_1 and t_2 be the intersection points, respectively. That is, T is a

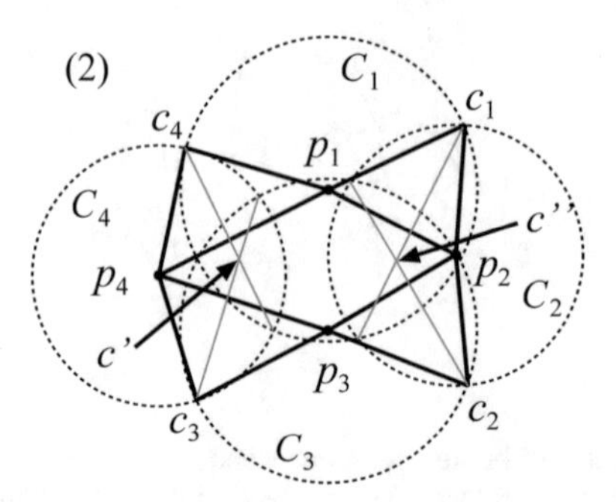

Fig. 7.8 (2) Glue c_3 and c_4 to obtain c', and glue c_1 and c_2 to obtain c''

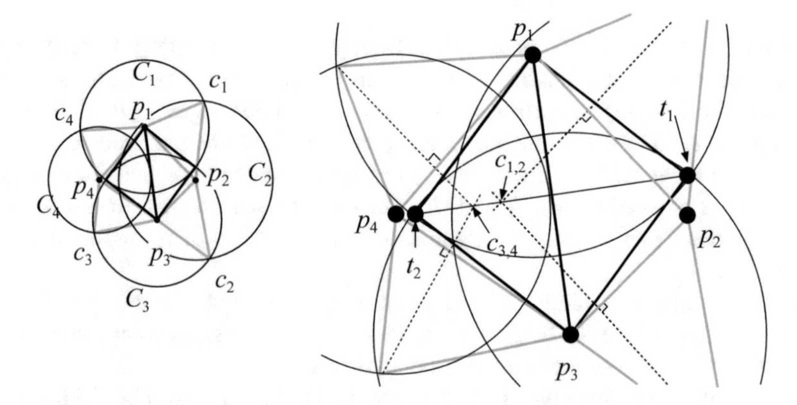

Fig. 7.9 (3) Folding along line segment p_1p_3 and its enlarged view

congruent triangle with $t_1p_1p_3$ and $t_2p_1p_3$, which intuitively stands up on p_1p_3 and has a structure that supports the apex of the folded bumpy pyramid from bottom.

We here consider four more intersection points: For each i, j, we denote by $c_{i,j}$ the intersection point of two line segments $\overline{c_i(0)c_i(1)}$ and $\overline{c_j(0)c_j(1)}$. By Theorem 7.5.1, two points $c_{1,2}$ and $c_{3,4}$ are on the line $\overline{t_1t_2}$.

If the point $c_{1,2}$ is closer to the point t_1 than the point $c_{3,4}$ as shown in the figure, to make a polyhedron by folding along the line p_1p_3, we have to mountain fold along it, make the point $c_{1,2}$ (made by gluing c_1 and c_2) and the point $c_{3,4}$ (made by gluing c_3 and c_4) closer, and glue them together. On the other hand, if the point $c_{3,4}$ is closer to the point t_1 than the point $c_{1,2}$, we have to make the point $c_{1,2}$ and the point $c_{3,4}$ farther by valley folding along the line p_1p_3; otherwise, we cannot fold into any polyhedron.

We next consider when we fold along the line p_2p_4 in the case that the point $c_{1,2}$ is closer to the point t_1 than the point $c_{3,4}$ in the above two cases. As in the above discussion, we consider the points $c_{2,3}$ and $c_{4,1}$, and let t_1' be the closer one of two intersection points of the circles C_2 and C_4. Then, we can see the case that the point $c_{1,2}$ is closer to the point t_1 than the point $c_{3,4}$ is the case that the point $c_{2,3}$ is closer to t'_1 than $c_{4,1}$. That is, as in the above analysis, we cannot fold into any polyhedron without valley folding along p_2p_4 to make the points $c_{2,3}$ and $c_{1,4}$ farther.

In the same way, if you cannot fold into a polyhedron unless you valley fold along the line p_1p_3, you cannot obtain a polyhedron unless you mountain fold along the line p_2p_4 in this case.

By the above discussion, when one diagonal line is selected and a convex pyramid is folded, then folding along the other diagonal line yields a concave pyramid. □

7.7 Folding Convex Pyramid

In this section, we consider the problem of folding a convex pyramid when we cannot fold a given petal polygon into a pyramid with flat base. As we have already seen in Theorem 7.6.1, when $n = 4$, there are three cases of folding bumpy pyramids; a petal polygon can fold into either no bumpy pyramid, one convex pyramid, or one convex and one concave pyramid. In this context, the convex pyramid has a special meaning from the others. One characteristic of a convex polyhedron is the Alexandrov's theorem. When the geometric structure is determined, the convex polyhedron is determined uniquely by Alexandrov's theorem. In the context of petal polygons, when convex pyramid can be folded from a petal polygon, its shape is uniquely determined. Actually, for general n, when the bumpy pyramids can be folded from a petal polygon, one of them is always convex, and all others are concave. It gives a somewhat strange impression. When n is large, the number of triangulations increases exponentially with respect to n. In general, some of them do not make any shape (the case (2) of Theorem 7.6.1 ($n = 4$) is a good example). When you can fold several polyhedra, only one of them is a convex polyhedron, and the rests are all concave. This is the purpose of this section to understand it. We prove Theorem 7.2.2 in this section. In order to do that, we first show the more specific Lemma 7.7.1.

Lemma 7.7.1 *Assume that a petal polygon consisting of* $2n$ *vertices* $P = (p_1, c_1, p_2, c_2, \ldots, p_n, c_n)$ *is given. If* P *can be folded into any bumpy pyramid, the unique convex pyramid can be folded. Furthermore, we can compute the crease lines for folding the convex pyramid in linear time* $O(n)$ *among the triangulations of the base* B.

The claim of Lemma 7.7.1 may be seemingly at a loss as to how to prove it. However, if you examine the analysis of the case of four points in Sect. 7.6, you will see a hint. This claim is closely related to the power diagram and it is not so difficult if you notice the relationship between the definition of power diagram and folding of petals. Specifically, there is a natural and beautiful correspondence between them, and when we use it, we can show Lemma 7.7.1. Let us first show the following theorem.

Theorem 7.7.2 *Let* $P = (p_1, c_1, p_2, c_2, \ldots, p_n, c_n)$ *be a petal polygon with a base* $B = (p_1, \ldots, p_n)$. *Define the weight at each point* p_i *as* $\ell_i = |p_ic_i|$ *and compute*

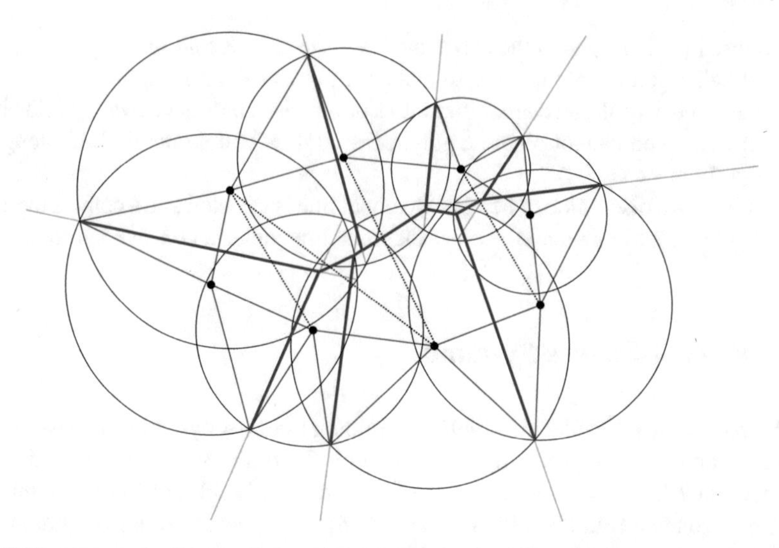

Fig. 7.10 An example of power diagram: Red lines are original base B and its surrounding petals. Draw a circle for each edge of petals as its radius and construct a power diagram based on them (blue lines). Then the projection of the vertex of a petal moves along the power diagram. From them, a triangulation of the base, that is, a crease pattern of the base is obtained (dotted line)

the power diagram G of B. Then, considering this power diagram as a graph and letting T be the triangulation of its dual, T is a set of diagonal lines of B. If P can be folded into a bumpy pyramid, when you fold the diagonal lines belonging to T in mountain and glue it in order given by the power diagram G, the convex pyramid can be folded.

Proof If a convex pyramid exists, its uniqueness is guaranteed by Alexandrov's theorem. Therefore, here we focus on the relevance to the power diagram and give a somewhat intuitive proof.

In Theorem 7.6.1, the circle C_i of the radius ℓ_i centered on each point p_i plays an important role. Considering two intersection points between C_i and C_{i+1}, one is $c_i(0)$, as we observed in Fig. 7.6, and the other is $c_i(1)$ where the point c_i moves to when we fold the flap to overlap the base B. Moreover, the line segment joining these two points is the projection of the trace of $c_i(t)$ when folding the flap. Considering this projection carefully, it is perpendicular to $p_i p_{i+1}$ that divides between the vertices p_i and p_{i+1} of the base B by the ratio of $\ell_i : \ell_{i+1}$. In other words, this line $c_i(0)c_i(1)$ corresponds to the power diagram of two points p_i and p_{i+1} on the plane of their weights given by ℓ_i and ℓ_{i+1}.

Now, since B is a convex polygon, the power diagram G of B is connected and acyclic by Lemma 7.3.2. An example is shown in Fig. 7.10. Some edges of the power diagram actually extend to infinity. Precisely, each edge of the power diagram that passes between the edge $p_i p_{i+1}$ of B corresponds to the "leaf" of a normal tree which is extending to infinity. On the other hand, the flap whose vertex is c_i is on the straight line extending to infinity and it moves from $c_i(0)$ to $c_i(1)$ when this flap is folded on B. Then projecting $c_i(t)$ onto the base plane B, the projected image forms the line segment $\overline{c_i(0)c_i(1)}$. Since we want to consider the trace of the projection of c_i here, we do not need line that follows infinity from $c_i(0)$. Therefore, for the sake of convenience, we consider the graph obtained by cutting off these lines as power diagram G. That is, G is a finite tree, and each leaf is associated with a certain vertex $c_i(0)$. For simplicity, we equate the leaves of G with the corresponding vertices c_i.

Here we consider the relationship between folding motion and G. Suppose that the edge connected to the leaf c_i and the edge connected to the leaf c_{i+1} are connected to an internal vertex w of G. (In the power diagram, such a pair of adjacent leaves always exists, but the proof is omitted here. As in Lemma 7.3.2, we can show from the fact that the internal vertex of G is degree 3.) Let us consider folding and gluing these two vertices c_i and c_{i+1}. Let $c_{i,i+1}$ be the resulting vertex made by gluing them. At this time, the vertex $c_{i,i+1}$ is created above the vertices p_i, p_{i+1}, p_{i+2} of B, and we can obtain a tetrahedron with these four vertices if we virtually glue a triangle $p_i p_{i+2} c_{i,i+1}$. We consider this triangle $p_i p_{i+2} c_{i,i+1}$ as a new flap. That is, we remove one triangle p_i, p_{i+1}, p_{i+2}, two flaps $p_i c_i p_{i+1}$, and $p_{i+1} c_{i+1} p_{i+2}$, and attach the triangle $p_i p_{i+2} c_{i,i+1}$ as a new flap. Then B is replaced by a new polygon B', which has one vertex fewer than B. Let P' be the new petal polygon obtained by this series of processing. Then, since all the processing so far have done locally, from now on, it is possible to handle the folding of P and the folding of P' in the same way. That is, the new flap $p_i p_{i+2} c_{i,i+1}$ in P' will be folded in mountain and glued with other flaps. Considering this fact in the context of the original P, after gluing the original two flaps $p_i c_i p_{i+1}$ and $p_{i+1} c_{i+1} p_{i+2}$, the virtual tetrahedron $c_{i,i+1} p_i p_{i+1} p_{i+2}$ is mountain folded along the line $p_i p_{i+2}$, and the virtual triangle $p_i p_{i+2} c_{i,i+1}$ will be glued with other flaps as a virtual flap. In other words, we can obtain a convex pyramid in a recursive way by folding mountain along the diagonal $p_i p_{i+2}$ above B.

Continuing the above discussion recursively, it is easy to see that you can glue in the following procedure: First, look at the internal vertex w which is connected to the two adjacent leaves c_i and c_{i+1} on G. Then glue the corresponding two flaps, remove them from P and B with the triangle $p_i p_{i+1} p_{i+2}$, and add the new flap $c_{i,i+1} p_i p_{i+2}$ to the new line segment $p_i p_{i+2}$ of B'. On the graph G, these operations correspond to remove the leaves c_i and c_{i+1} and we consider w as a new leaf of the new tree G'. We repeat the same procedure.

Here, considering the triangulation T of B as a dual of G, a choice of a new internal vertex w in the above procedure corresponds to choose an edge $p_i p_{i+2}$ of T. Therefore, it is now obvious that this triangulation T gives us diagonal lines to fold the convex pyramid. □

In order to complete the proof, it remains to show that the power diagram can be computed in linear time.

Lemma 7.7.3 *The power diagram of a vertex set given by the base convex n-gon B can be computed in $O(n)$ time.*

Proof Here we point out only the fact that Aggarwal et al. developed a linear time algorithm for this problem in 1989 in [AGS+89], and we omit the proof. Their algorithm is based on a kind of divide-and-conquer method, but it is rather skillful. □

Incidentally, a Voronoi diagram is a special (or simple) notion of the power diagram, but the algorithm for computing the Voronoi diagram for a given point set in $O(n \log n)$ time is not that simple. There are several algorithms for computing the Voronoi diagram in $O(n \log n)$ time with a long history, and they are not straightforward. As with sorting algorithms, there are basic and various approaches, and not that easy. (It is similar in that it is not a very difficult problem unless you want to make computation time so fast.) By using the fact that the point set is in convex position in this problem, we can improve the running time to linear time, but it is the core part of computational geometry and it goes beyond the scope of this book. Interested readers should refer to authentic textbooks of computational geometry, e.g., [BCK+10].

7.8 Bumpy Pyramid of the Maximum Volume

Finally, we introduce an algorithm to compute the bumpy pyramid of the maximum volume. In most cases, the convex pyramid appears to be the maximum volume; however, as shown in Fig. 7.3, the concave pyramid may have a larger volume than the convex pyramid in some cases. We here note that, if one triangulation is given, we can compute the volume by using the result of Sabitov [Sab98]:

Lemma 7.8.1 *When the petal polygon P and a triangulation T of its base B are given, the volume of the bumpy pyramid folded from them can be computed in linear time.*

Proof As shown in the proof of Theorem 7.3.1, T has at least two ears. Let p_{i-1}, p_i, p_{i+1} be the vertices of one of these ears. Then, when P is folded along T, the vertices $cp_{i-1}p_i p_{i+1}$ form a tetrahedron. Since we know all lengths of six edges, the volume of this tetrahedron can be computed in a constant time (using the formula given in Column 7.6).

Similar to the proof of Theorem 7.7.2, we remove two flap triangles $p_{i-1}p_ic_{i-1}$ and $p_ip_{i+1}c_i$ from P and remove the triangle $p_{i-1}p_ip_{i+1}$ from B. Then we obtain a new petal polygon P' by replacing them by a new flap triangle $p_{i-1}p_{i+1}c_{i,i+1}$ and its triangulation T' obtained from T by removing the line segment $\overline{p_{i-1}p_{i+1}}$. For these P' and T', we again consider a tetrahedron on one of the ears of T', obtain its volume, and repeat in the same way. When you add all the volumes of the series of tetrahedra obtained in this way, it will be the volume of the completed bumpy pyramid. □

However, the number of triangulations of a convex n-gon is exponential (precisely, it is $\frac{1}{n-1}\binom{2n-4}{n-2}$); therefore, trying all cases is not efficient. In fact, there is a useful method that can be used in common for the problem to find the triangulation with the optimal value among such exponential triangulations. We show it by the following theorem.

Theorem 7.8.2 *Assume that a petal polygon $P = (p_1, c_1, p_2, c_2, \ldots, p_n, c_n)$ of $2n$ vertices is given. Let $B = (p_1, p_2, \ldots, p_n)$ be the base of P. Then, we can compute the crease lines of B (namely, triangulation of B) that maximize the volume in $O(n^3)$ time.*

Proof We use a *dynamic programming* for the triangulation.[6]

Here we define a subproblem $S(i, k)$ for a partial bumpy pyramid using vertices from the vertex i to the vertex $i + k$. (Hereafter, we will assume that $1 \leq i \leq k < n$ without loss of generality. If the index exceeds n, return the index to 1.) For this subproblem, let $w(i, k)$ be the maximum volume among one of the bumpy pyramids that can be folded only by this partial petal polygon. This subproblem can be considered for computing the maximum volume of bumpy pyramids folded from a petal polygon P' that consists of a part of P at the vertices $p_i, c_i, \ldots, c_{i+k-1}, p_{i+k}$, and one additional triangular flip $c' p_i p_{i+k}$, where c' is the vertex taken to satisfy $|c' p_i| = \ell_i$ and $|c' p_{i+1}| = \ell_{i+1}$. Here, we consider $w(i, 1) = 0$ when $k = 1$ for any i since it can be seen as gluing two congruent flaps. Since P' is P itself when $k = n - 1$, we have to compute $w(1, n - 1)$ to solve the original problem.

As is often the case with dynamic programming on triangulation, if you divide the problem into two subproblems with a diagonal of B, the size of the table you manage will become exponential and it will not be efficient eventually. The key point is to divide the problem into subproblems with a triangle.

For the 3-tuple (i, j, k) of integers, we consider forming the tetrahedron using the triangle $\triangle p_i p_{i+j} p_{i+k}$ of three vertices of B and three associated lengths ℓ_i, ℓ_j, ℓ_k. Let $V(i, j, k)$ be the volume of this tetrahedron. This can be computed by the equation shown in Column 7.6. When the obtained $V(i, j, k)$ satisfies $V(i, j, k) \geq 0$, that is, when a tetrahedron is established, we call this 3-tuple *valid*. Then, $w(i, k)$ satisfies the following formula for k with $1 < k < n$:

$$w(i, k) = \begin{cases} \max_{1 \leq j < k} V(i, j, k) + w(i, j) + w(i + j, k) & \text{when } (i, j, k) \text{ is valid} \\ -\infty & \text{otherwise} \end{cases}.$$

Therefore, if you compute $w(i, k)$ for each $k = 1, 2, \ldots, n - 1$ in ascending order for each vertex $i = 1, 2, \ldots, n$, you can solve the original problem in $O(n^3)$ time in the standard dynamic programming. Moreover, when solving each subproblem, if you store the index j giving the maximum value, you can restore the crease line (one of the triangulations of B) that maximizes the volume. □

---

[6]Dynamic programming is one of techniques of algorithms. It is briefly described as follows. We first define a solution in a recursive way. Then, instead of recursive calls, the algorithm constructs the solution in a table in the bottom-up manner.

7.9 Unsolved Problems

The bumpy pyramid folding problem dealt with in this section turned out to be solvable efficiently; however, it does not solve the problem for general convex polyhedra, and handles a rather special polyhedra. It is considered to be a very promising problem to develop algorithms to fill this gap so far. A bumpy pyramid itself is a very special shape, but considering only the vicinity of the apex of a general convex polyhedron, it may be possible to consider the same problem as this. In other words, the technique in this section may be used when locally solving the net of a general convex polyhedron.

Looking ahead to such problems, the following bumpy pyramid problem has not been yet solved.

Open Problem 7.9.1 *For the bumpy pyramid problem, what is the condition for the convex polyhedron to have the maximum volume? On the contrary, the condition for a volume maximum of a non-convex polyhedron can be the problem.*

None of the necessary and sufficient conditions are known at all. Intuitively, somehow, other than "extremely" and "distorted", the convex one seems to have the maximum volume. However, giving concrete characterizations such as "extreme" and "distorted" is a remaining task.

The bumpy pyramids dealt with in this section, although it is bumpy, does not cover all possible cases. For example, we do not consider the case where the apex vertex becomes a saddle point exceeding 360°. We also do not consider shapes in which the base B is folded back in a zigzag manner, resulting in a part of the flap to turn over and the polyhedron gets inside. The general theory which also applies to such polyhedra is still under development.

References

[AGS+89] A. Aggarwal, L.J. Guibas, J. Saxe, P.W. Shor, A linear-time algorithm for computing the voronoi diagram of a convex polygon. Discret. Comput. Geom. **4**(1), 591–604 (1989)

[Aur87] F. Aurenhammer, Power diagrams: properties, algorithms and applications. SIAM J. Comput. **16**, 78–96 (1987)

[BCK+10] M. de Berg, O. Cheong, M. van Kreveld, M. Overmars, *Computational Geometry: Algorithms and Applications* (Springer, Berlin, 2010)

[BI08] A.I. Bobenko, I. Izmestiev, Alexandrov's theorem, weighted Delaunay triangulations, and mixed volumes (2008), arXiv:math.DG/0609447

[DO07] E.D. Demaine, J. O'Rourke, *Geometric Folding Algorithms: Linkages, Origami* (Cambridge University Press, Polyhedra, 2007)

[Die96] R. Diestel, *Graph Theory* (Springer, Berlin, 1996)

[KPD09] D. Kane, G.N. Price, E.D. Demaine, A pseudopolynomial algorithm for Alexandrov's theorem, in *11th Algorithms and Data Structures Symposium (WADS 2009)*, pp. 435–446. Lecture Notes in Computer Science vol. 5664 (Springer, 2009)

[Sab98] I.K. Sabitov, The volume as a metric invariant of polyhedra. Discret. Comput. Geom. **20**, 405–425 (1998)

Chapter 8
Zipper-Unfolding

Abstract In this chapter, we focus on the edge-unfolding of convex polyhedron. It is conjectured that we can always do that, however, it is not yet settled. Thus we focus on the unfolding that is realized by zipper.

8.1 Unfolding Along Hamilton Path

As mentioned in Chap. 1, it is conjectured that any convex polyhedron can be unfolded into a net (without overlap) by cutting along edges properly. This is an unsolved problem that has not been solved for hundreds of years, and it is unlikely to be solved soon. One of the reasons for this difficulty is that the edge-unfoldings have two aspects.

First, it has a graph-theoretic property when the vertices and edges of the convex polyhedron are regarded as contents of a graph. Any edge-unfolding considered on this graph is always a spanning tree of the graph. In other words, looking at the prospect from the aspect of graph theory, the conjecture can be regarded as a problem for finding a "good" spanning tree with some certain indicators. In general, it is easy to find one arbitrary spanning tree for a given graph; however, finding a "good" spanning tree can be difficult in general. Moreover, in the current problem, it is not clear what to use as indicators. This is one aspect that gives difficulties.

The other is the geometric aspect of edge-unfolding. After all, when a convex polyhedron and its edges to be cut are given, it is a very complicated geometric problem to find how each surface is developed, and as result, how all the faces are arranged on a plane.

Therefore, in order to solve the conjecture affirmatively, for any given convex polyhedron, it is necessary to always find a spanning tree with good properties in the sense that all the faces are successfully and geometrically unfolded without any overlapping.

It is a natural idea to tackle a limited problem with some restrictions as one approach to the problem having both geometric and graph-theoretic difficulties. If we expand the range to be solved little by little, one day we may find a general way

R. Uehara, *Introduction to Computational Origami*,
https://doi.org/10.1007/978-981-15-4470-5_8

to compute an edge-unfolding of any convex polyhedron, or conversely, we may find a convex polyhedron that always overlaps for any edge-unfolding.

In this chapter, as the forefront of this research topic, let us first introduce convex polyhedra that we know edge-unfolding can be done. By adding geometric restrictions to polyhedra, it is guaranteed that these convex polyhedra always have spanning trees with good properties. This may be unexpected, but it is known only when it is quite limited. Next, we consider a special development method called *Hamiltonian unfolding*, which is also known as *zipper-unfolding*. This is an unfolding method that restricts the spanning tree to a *path* instead of a *tree*. In other words, there is no branch in the route. Intuitively, once you insert scissors at a certain vertex, you cut through along consecutive line segments, and cut the whole polyhedron by this stroke. We show some known results for this special way of edge-unfolding.

8.2 Convex Polyhedra that Can Be Unfolded by Edge-Unfolding

First, we introduce convex polyhedra that known to be unfoldable by edge-unfoldings. A *dome* has the base B, and all other faces share an edge with B. Intuitively, it is composed only of the base and the side faces, and it has only vertices on the top. A *prismoid* is a polyhedron with a top T and a base B where T and B are polygons with the same number of vertices such that the corresponding angles are equal and the corresponding edges are parallel. By this definition, we can see that all sides are trapezoids. Convex domes and convex prismoids are known to be always unfolded by edge-unfoldings. The known way of unfolding is called *volcano-unfolding*. Intuitively, it unfolds so that each side is attached to the bottom. In the case of a prismoid, the top T is put on some side. It may look easy, but it is not that easy to guarantee that the sides do not overlap and that the sides can be unfolded so that the top will not overlap somewhere at *any* time. For a detailed discussion, please refer to [DO07].

A *prismatoid* is known as a polyhedron that forms a superset of prismoids. If you have two polygons as the top T and the base B in parallel, their convex hull forms a prismatoid. Here T and B do not necessarily have the same number of vertices, and the corresponding angles may be different. Each side of a prismatoid is always a triangle or a trapezoid. The smallest class among the convex polyhedra which are not known whether it is always unfoldable by edge-unfolding is the prismatoids.

Open Problem 8.2.1 *Is any convex prismatoid unfolded to a net by edge-unfolding?*

8.3 Zipper-Unfolding

We next consider the case where the lines to be cut are restricted. We restrict the spanning tree by limiting a “tree” to be a “path”. By the property of the unfolding, this path visits every vertex on the polyhedron exactly once. In terms of graph theory,

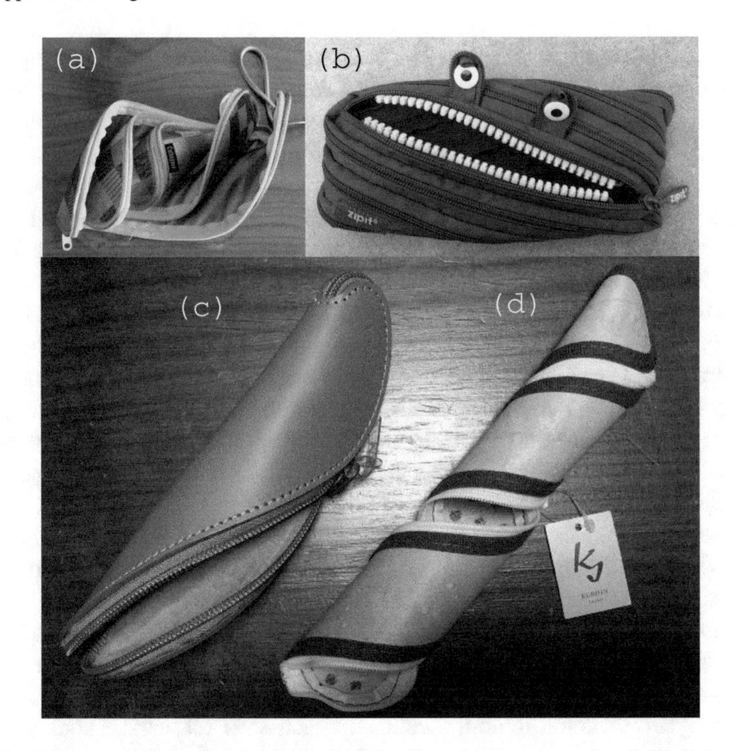

Fig. 8.1 Various containers that can be unfolded with a zipper: **a** is made of silk, **b** is made of plastic, **c** is made of leather, and **d** is wooden

this is a *Hamilton path*. Therefore, this method of unfolding is sometimes called *Hamiltonian unfolding* [DDL+10]. From an industrial point of view, it can be realized with a zipper, and there are actually some products made with this idea (Fig. 8.1). Thus, we call it *zipper-unfolding* instead of Hamiltonian unfolding in this book.

From a graph-theoretical point of view, the zipper-unfolding problem is closely related to the Hamilton path problem on the graph induced by the vertices and edges on the polyhedron. That is, if the polyhedron can be unfolded by zipper-unfolding, the corresponding graph trivially has a Hamilton path. On the other hand, just because it has a Hamilton path, it does not necessarily have a net by zipper-unfolding. Paper may overlap when we unfold it by zipper-unfolding. Demaine et al., who introduced this notion in 2010, investigated the existence of zipper-unfoldings for various concrete polyhedra such as regular polyhedra and quasi-regular polyhedra [DDL+10].

In the paper, they showed that *rhombic dodecahedron* had no nets by zipper-unfolding. However, the reason was that the graph induced by the rhombic dodecahedron had no Hamilton path in the first place. It was also known that there were several other polyhedra without nets by zipper-unfoldings, but that was all because the corresponding graphs do not have Hamilton paths.

However, this research story has gone wrong direction. Initially, in investigation of the difficulty of edge-unfolding of convex polyhedra, it should have been an approach to study the other by limiting one of the two aspects. In other words, given the origin that the first motivation restricted to zipper-unfolding was the constraint introduced to study the geometric property of overlapping faces on a plane, the argument that "it cannot be unfolded by zipper-unfolding because the graph does not have a Hamilton path" seems to be a misdirected route to the correct object. There is doubt as to whether it is reasonable to restrict zipper-unfolding in the first place.

With this background in mind, we have reached the question of whether there is a convex polyhedron with a *geometric* property that the graph induced by the convex polyhedron has Hamilton paths and it always overlaps with any zipper-unfolding. Roughly speaking, is there a frustrating polyhedron such that there are many ways to unfold by zipper-unfolding, but no matter how you open up, it does not work as the faces always overlap? The answer is "YES". Let us introduce convex polyhedra with this interesting property.

The series of convex polyhedra shown here are simple domes, and it is easy to unfold without overlapping by volcano-unfolding as shown in Sect. 8.2. However, if you restricted yourself to zipper-unfolding, interestingly, overlapping will occur no matter how you unfold. On the other hand, there are exponentially many Hamilton paths on this dome. This example implies that we cannot avoid geometric discussion even if we attempt to discuss the existence of unfolding only limited to a very simple convex polyhedron since graph-theoretic approach does not work in some cases.

8.3.1 Convex Polyhedra Which Cannot Be Unfolded by Zipper-Unfolding

We here consider the dome introduced in Sect. 8.2. It is a rather simple convex polyhedron consisting of sides sharing only one edge with the base B. However, if we design it appropriately, while it is possible to construct interesting domes that have exponentially many Hamilton paths, any zipper-unfolding always overlaps. Before showing a concrete dome, we show one geometric lemma.

Lemma 8.3.1 *Let $\theta(n) = \frac{2\pi}{n}$ for a positive integer n. Also, let $T(n)$ be an isosceles triangle whose angle at the apex is $\theta(n)$. We assume that the length of two edges of T is unit length. Now we prepare eight copies of $T(n)$ and arrange them as shown in Fig. 8.2. Here, each bold line is shared by two triangles. Then, if $n > 12$, the triangles T_4 and T_8 overlap.*

Intuitively, it is almost trivial that triangles T_4 and T_8 overlap if n is large enough, but we compute the exact value to make sure.

Proof To compute lengths, we suppose that the apex of the triangle T_5 is on the origin $O = (0, 0)$, and the line connecting the apices of T_1 and T_5 is on the y-axis.

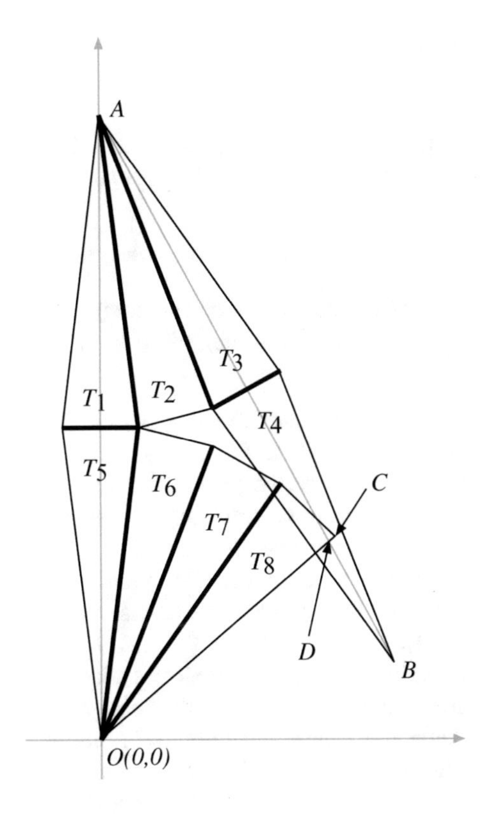

Fig. 8.2 Overlapping triangles

Let A and B be the apices of T_1 and T_4, respectively. Then, each coordinate can be obtained as follows:

$$A = \left(0, 2\sin\frac{\theta(n)}{2}\right),\ B = \left(2\sin\frac{\theta(n)}{2}\sin 2\theta(n), 2\sin\frac{\theta(n)}{2}(1-\cos 2\theta(n))\right).$$

On the other hand, let C be the one of two base angles of T_8 farther from T_5. Then its coordinate is computed as follows:

$$C = \left(\cos\frac{7\theta(n)}{2}, \sin\frac{7\theta(n)}{2}\right).$$

Now we consider the intersection D of lines AB and OC. (Precisely, D is the intersection of two lines containing the line segment AB and the line segment OC.) Then the point D is the inside of two triangles T_4 and T_8 when $|OD| < 1$. By simple calculation, we obtain

$$D = \left(\frac{2\sin\frac{\theta(n)}{2}}{\cot 2\theta(n) + \cot\frac{7\theta(n)}{2}}, \frac{2\sin\frac{\theta(n)}{2}\tan 2\theta(n)}{\tan 2\theta(n) + \tan\frac{7\theta(n)}{2}}\right).$$

Therefore, we have

$$|OD|^2 = 4\sin^2\frac{\theta(n)}{2}\left(\frac{1}{\left((\cot 2\theta(n) + \cot\frac{7\theta(n)}{2}\right)^2} + \frac{\tan^2 2\theta(n)}{(\tan 2\theta(n) + \tan\frac{7\theta(n)}{2})^2}.\right)$$

Since this value is less than 1 when $n > 12$, T_4 and T_8 overlap each other then. □

Theorem 8.3.2 *There are infinitely many domes that cannot be unfolded without overlap in any zipper-unfolding.*

Proof For any positive integer n, construct a dome $D(n)$ as follows. The base $B(N)$ is a regular $2n$-gon. Let $p_1, p_2, \ldots, p_{2n}$ be the vertices of $B(n)$. This dome $D(n)$ has the apex c on the perpendicular set at the center of $B(n)$. The height of this apex c is assumed to be extremely low. Then we draw a small circle C centered at the apex c and take n points, $q_1, q_2, \ldots, q_n$ at equally spaced intervals on this C. That is, these n points form a small regular n-gon centered on c. For simplicity, we assume that the height of c and the radius of C are almost 0. Now we add edges $\{p_{2i-1}, q_i\}$ and $\{p_{2i}, q_i\}$ for each $i = 1, 2, \ldots, n$. We rotate circle C appropriately to make each triangle $q_i p_{2i-1} p_{2i}$ an isosceles triangle. For each $i = 1, 2, \ldots, n$, we add edges connecting the apex c and the point q_i. As an example, we show the top view of the dome $D(n)$ for $n = 3$ in Fig. 8.3. Intuitively, around the bottom B, the n isosceles triangles and the similar n pentagons are alternately lined and surrounded. We show that although there are many Hamilton cycles on this dome $D(n)$, we cannot unfold by zipper-unfolding when $n > 12$.

We assume that we can obtain a net of $D(n)$ by cutting along a Hamilton path P. For each vertex v on the path P, we denote the number of edges of P connected to v by $\deg_P(v)$. Since P is a Hamilton path, $\deg_P(v) = 1$ for the two endpoints v and $\deg_P(u) = 2$ for all other vertices u. Therefore, $\deg_P(c)$ is 1 or 2, and $\deg_P(q_i) = 2$ at almost all vertices q_i. By this fact, we can observe that the path (p_{2i-1}, q_i, p_{2i})

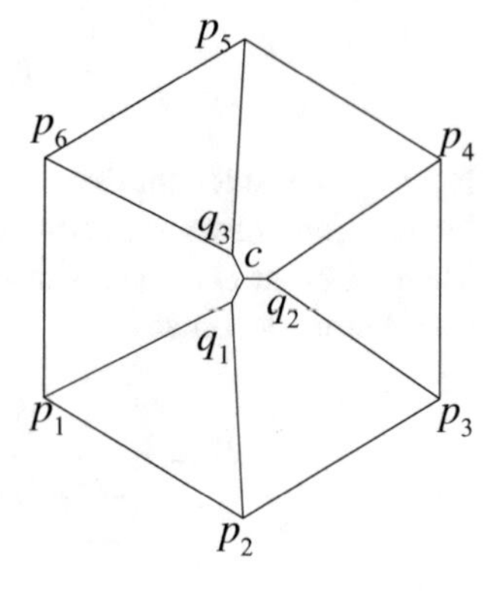

Fig. 8.3 Top view of dome $D(3)$

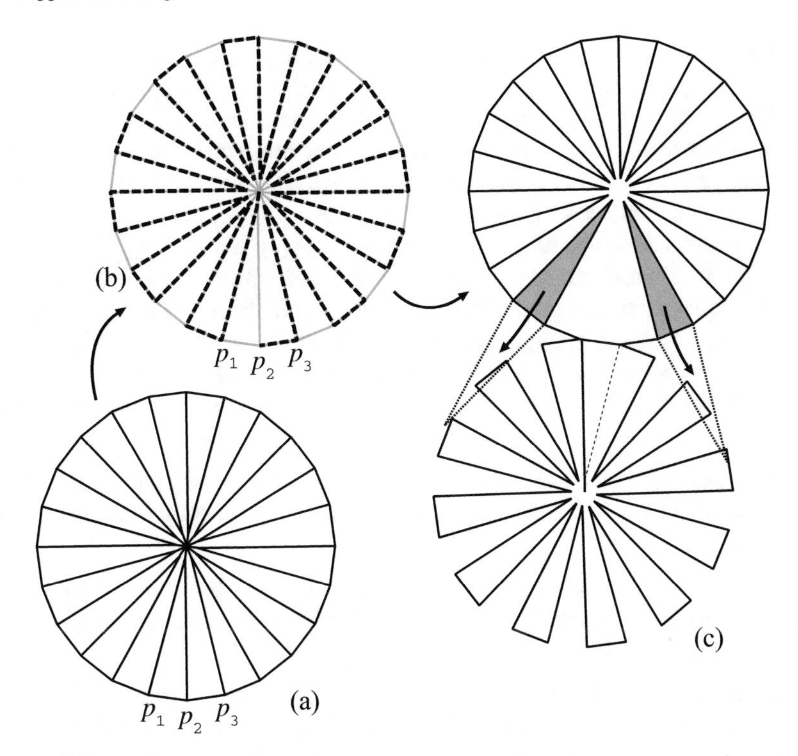

Fig. 8.4 One of the ways of zipper-unfolding of $D(12)$

is a part of P for almost all q_i. That is, almost all the isosceles triangles are flipped along the edge sharing with the base like flower petals.

We here consider two cases. First case is that c is the endpoint of P. Without loss of generality, we can assume that (c, q_1, p_1) is the end of the path P. Then c is not cut except along the edge (c, q_1), and hence P contains (p_{2i-1}, q_i, p_{2i}) as a subpath for all i with $1 < i \leq n$ (i.e., except $i = 1$). Then there are only two ways to construct a Hamilton path: One is $(c, q_1, p_1, p_{2n}, q_n, p_{2n-1}, \ldots, p_4, q_2, p_3, p_2)$, and the other is $(c, q_1, p_1, p_2, p_3, q_2, p_4, \ldots, p_{2n-1}, q_n, p_{2n})$.

In the first case, the situation is as shown in Fig. 8.4b and c: First, we cut along the dotted line in Fig. 8.4b. Next, we turn over the part of the lid, that is, the triangle $p_1 p_2 q_1$ and all pentagons (Fig. 8.4c). Then, we have to open the remaining triangles; however, the gray triangle overlaps with the other parts of lid by Lemma 8.3.1 when the circle C is sufficiently small, the dome height is sufficiently low, and $n > 12$. That is, in this case, the dome cannot be unfolded without overlapping. The second case is easier; the triangles closest to the lid will overlap again when we unfold it. Therefore, when c is the end of a Hamilton path P, an overlap occurs in any zipper-unfolding.

Next, we consider the case where c is not the endpoint of P. Without loss of generality, we assume that (q_i, c, q_1, p_1) is a subpath of P for some i. When q_i is an endpoint, the same argument can be used as in the first case: If $q_i = q_2$ or $q_i = q_n$, either one of the two triangles will be overlapped with the lid part, and both of two triangles will be overlapped otherwise. Hence, we consider the case where a part of the path contains (p_j, q_i, c, q_1, p_1) for some j. Now we have $j = 2i - 1$ or $j = 2i$. Then, when we remove all vertices $\{p_j, q_i, c, q_1, p_1\}$ from the graph induced by the vertices and edges of the current $D(n)$, the graph is divided into two subgraphs. Therefore, we call the graph induced by the vertex set $\{p_2, p_3, \ldots, p_{j-1}, q_2, q_3, \ldots, q_{i-1}\}$ the *right* graph, and the graph induced by the other $\{p_{j+1}, p_{j+2}, \ldots, p_{2n}, q_{i+1}, \ldots, q_n\}$ the *left* graph. Then P can be divided into three parts, which are the path P_r visiting all the vertices of the right graph, the path P_l visiting all the vertices of the left graph, and the path (p_j, q_i, c, q_1, p_1) connecting the two paths P_r and P_l. Here, let P' be the longer path between P_r and P_l. Then we can apply the same argument on the path obtained by joining P' and (p_j, q_i, c, q_1, p_1), and we can show that overlap occurs. □

In a zipper-unfolding, the degree of each vertex in the cutting path is 1 or 2. The discussion above can be extended to a general tree of the maximum degree at most $k \geq 2$.

Theorem 8.3.3 *We fix any positive integer $k \geq 2$. Then there are infinitely many domes which cannot be unfolded without an overlap by any cutting tree of the maximum degree at most k.*

We here note that, in the dome $D(n)$, the degree of any vertex is 3 except for the central vertex c. In other words, in a tree whose maximum degree is limited in Theorem 8.3.3, the only vertex of degree greater than 3 is the central vertex.

Proof We consider a dome $D(n)$ with $n > 6k$ for a given k. Let T be any spanning tree of the graph induced by the vertices and edges of $D(n)$ such that the maximum degree of T is at most k. We prove that an overlap always occurs when unfolding $D(n)$ by cutting along the edge of T. By definition, the degree of the central vertex c in T is at most k. Let $N_T(v)$ be the set of neighbors of v on T, and we define $N_T(N_T(c)) = \cup_{q \in N_T(c)} N_T(q)$. Furthermore, let T_c be the subtree of T induced by the vertex set $\{c\} \cup N_T(c) \cup N_T(N_T(c))$. Then since each q_i can supply leaves only from p_{2i-1} and p_{2i}, T_c has at most $2k$ leaves. Here, by a theorem on average value,[1] there is always a pair of two leaves p and p' of T_c such that there are at least $(n - k)/k > 5$ consecutive triangles on the bottom boundary between them (Fig. 8.5). When we unfold the dome along the edge of T, these consecutive triangles behave the same as we have already seen in the proof of Theorem 8.3.2. More precisely, the set of pentagons between p and p' constitutes a lid, and the whole is flipped along an edge $\{q, q'\}$ on the boundary of the base B (Fig. 8.5). When all the triangles between p and p' are flipped, two triangles (gray triangles in Fig. 8.5) sharing points q and q' with lid overlap the lid by Lemma 8.3.1. □

The dome $D(n)$ dealt with in Theorems 8.3.3 and 8.3.2 has only polynomial number of Hamilton paths for n. However, this number can be increased exponentially.

[1]This is a common theorem that there is always a person who marks a score higher or equal to the average score. It is intuitively trivial, but very powerful theorem when we show existence.

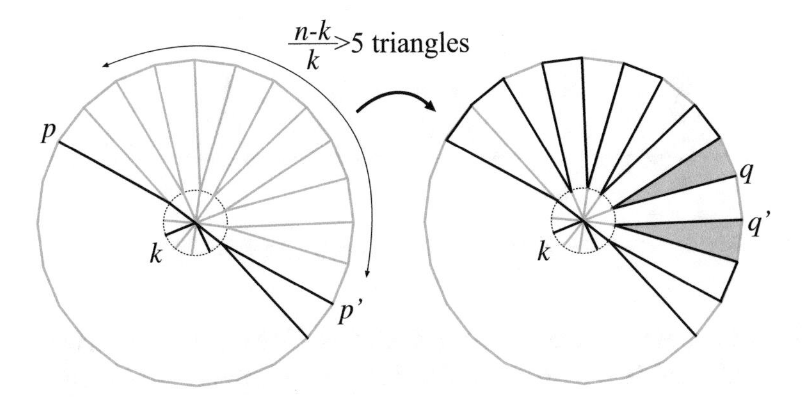

Fig. 8.5 When the maximum degree of a tree is bounded

Fig. 8.6 Top view of $D'(3)$

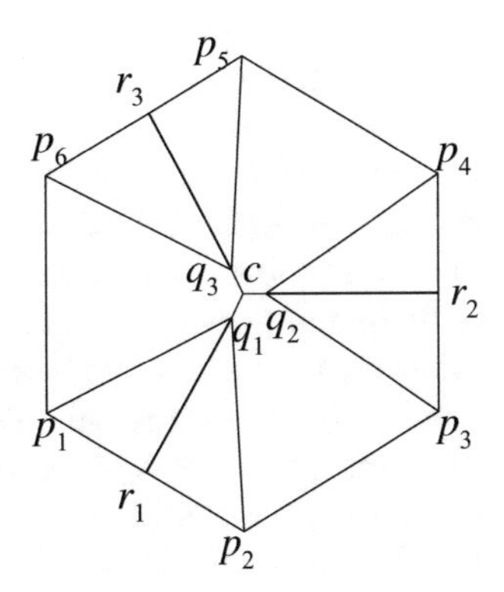

Corollary 8.3.4 *We can extend Theorems 8.3.3 and 8.3.2 to the domes having exponential number of Hamilton paths with respect to the number of vertices.*

Proof For a dome $D(n)$, we divide each isosceles triangle $p_{2i-1}q_i p_{2i}$ into two triangles $p_{2i-1}q_i r_i$ and $p_{2i}q_i r_i$, where r_i is the middle point of the edge $p_{2i-1}p_{2i}$ (Fig. 8.6). As shown in the proof of Theorem 8.3.3, almost all isosceles triangles $p_{2i-1}q_i p_{2i}$ are cut along (p_{2i-1}, q_i, p_{2i}) and flipped. This operation does not change even after dividing. However, when we try to construct a zipper-unfolding to pass through the new four vertices $\{p_{2i-1}, q_i, r_i, p_{2i}\}$, we will have two choices around the vertices. More specifically, we cut along either the path $(p_{2i-1}, q_i, r_i, p_{2i})$ or the path $(p_{2i-1}, r_i, q_i, p_{2i})$. Thus $D'(n)$ has exponentially many Hamilton paths. When n is sufficiently large, we can see that the arguments of Theorems 8.3.3 and 8.3.2 also hold for $D'(n)$, and we have an overlap for any zipper-unfolding of $D'(n)$. □

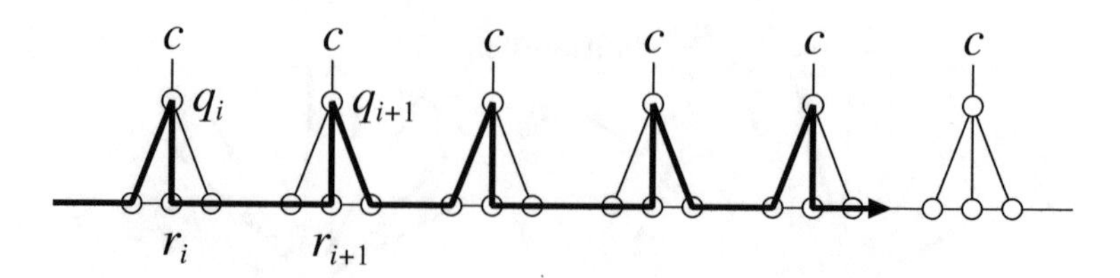

Fig. 8.7 Side view of a dome $D'(n)$ that has exponentially many Hamilton paths

8.4 Summary of the Current States of Convex Polyhedron Unfoldable by Edge-Unfolding and Zipper-Unfolding

For the unresolved open problem that asks whether any convex polyhedron can be unfolded along cutting its edges without overlap or not, we see that very weak results are currently known. Let us list some unresolved problems, in particular, that may be solved in a while.

As we saw in Sect. 8.2, any dome can be unfolded in volcano-unfolding. However, as shown in Sect. 8.3.1, it may not be unfolded by any zipper-unfolding. On the other hand, it is known that prismoids can be unfolded by an edge-unfolding by volcano-unfolding like dome. Then how about zipper-unfolding? From around here, what is known is somewhat complicated. It is known that if a prismoid is a *nested prismoid* which satisfies the special condition, it has a zipper-unfolding without an overlap. Since the discussion is a bit complicated, it is omitted here, but the intuitive definition of a nested prismoid is not so difficult. A prismoid has parallel lid and bottom. A prismoid is *nested* if the lid is completely contained in the bottom when looking at them from its top. Formally, a prismoid is nested if the projection of the lid is completely contained in the bottom. It is known that any nested prismoid can be unfolded to a net by zipper-unfolding. Intuitively, we can unfold by spreading the side of the nested prismoid like ribbon, and place the top lid and the bottom sole on each side without overlap (see [DDU13] for further details). Then what about a general prismoid? This is unresolved. As a partial solution, the following two results are known.

First, it is shown that there are only a polynomial number of Hamilton paths in a prismoid. Therefore, in contrast to the dome shown in Corollary 8.3.4, by generating all possible Hamilton paths and examining them one by one, we can check if a given prismoid can be unfolded by zipper-unfolding in a polynomial time. In other words, from an algorithmic point of view, it is a tractable problem.

Second, the problem of unfolding of a prismoid without its upper lid and the lower base has been researched. That is, it has only sides and ribbon-like shape. To put it a little more specifically, the following problem has been solved: "Is it possible to unfold a prismoid without having an overlap by removing the lid and the bottom from it and cutting along one edge of the remaining sides?" Just to answer this question, it is always [Yes]. Intuitively, this may seem obvious, but in fact it is a very complicated problem. The proof of the fact that the answer to this question is [Yes] for the sides of a nested prismoid is the topic of a journal paper [ADL+08], and the proof for the sides of a general prismoid is a doctoral thesis [Alo05] of Aloupis who is one of the authors of the paper [ADL+08] for a nested prismoid. In other words, looking only at the conclusion, no matter what kind of sides of a prismoid; if you cut a *proper* edge, it will be unfolded without any overlap. Therefore, similar to the nested prismoid, if we prove that the top lid and the bottom base can be successfully attached so that they do not overlap anywhere after cutting and unfolded on the sides, we can prove that any prismoid can be unfolded by zipper-unfolding; however, this last part is still unsolved.[2] In summary, the following is the unsolved problem of the forefront concerning zipper-unfolding.

Open Problem 8.4.1 *Can any prismoid be unfolded without overlap by zipper-unfolding?*

References

[DO07] E.D. Demaine, J. O'Rourke, *Geometric Folding Algorithms: Linkages, Origami* (Cambridge University Press, Polyhedra, 2007)

[DDL+10] E.D. Demaine, M.L. Demaine, A. Lubiw, A. Shallit, J.L. Shallit, Zipper unfoldings of polyhedral complexes, in *Canadian Conference on Computational Geometry (CCCG 2010)* (2010), pp. 219–222

[ADL+08] G. Aloupis, E.D. Demaine, S. Langerman, P. Morin, J. O'Rourke, I. Streinu, G. Toussaint, Edge-unfolding nested polyhedral bands. Comput. Geometry **39**, 30–42 (2008)

[Alo05] G. Aloupis, Reconfigurations of polygonal structure. Ph.D. thesis, School of Computer Science, McGill University (2005)

[DDU13] E.D. Demaine, M.L. Demaine, R. Uehara, Zipper unfoldability of domes and prismoids, in *Canadian Conference on Computational Geometry (CCCG 2013)* (2013), pp. 43–48

[2] I have asked directly to Dr. Greg Aloupis, the author of [Alo05], but his answer was "It is complicated".

Chapter 9
Rep-Cube

In this chapter, we introduce a new concept of rep-cube and its known results. It is a fledgling concept born in 2016, so there are many topics to be studied.

9.1 History and Preparation of the Rep-Cube

First, let's introduce the history of the rep-cube very briefly. The first appear is the polyomino introduced in Sect. 3.1. This is Solomon W. Golomb devised and studied in 1954 [Gol96]. Since then, it has been very widely studied in the puzzle society [Gar08]. For example, Fig. 82 of the book [Gar08] shows a method of covering a cube of size $\sqrt{10} \times \sqrt{10} \times \sqrt{10}$ with all 12 types of pentomino (polyomino of area 5).

In 1962, Golomb devised another interesting concept, the rep-tile. Dividing a polygon into k pieces, if it can all be congruent and similar to the original polygon, then this polygon is called *rep-tile* of order k.[1] For details, refer to the book [Gar14, Chap. 19].

Then the story flies in 2016. Martin L. Demaine advocated the following question at a workshop where I often participate: "Is there a polyomino that is a net of a cube, which can be divided into k polyominoes, each of which is a net of a cube?" Several solutions and related results were found at the workshop, and we wrote a paper [ABD+17] that was so interesting. In that time, I named it *rep-cube* because it resembles the two famous concepts by Golomb.[2]

[1] Note that "rep-tile" is the word play of "reptile".

[2] In fact, Solomon W. Golomn was born in 1932 and passed away in 2016. We gave the name to mourn it. Later, earlier research in the puzzle society was found the same concept as rep-cube. See [Tor92, Tor02, Tor02b, Tor03] for further details.

R. Uehara, *Introduction to Computational Origami*,
https://doi.org/10.1007/978-981-15-4470-5_9

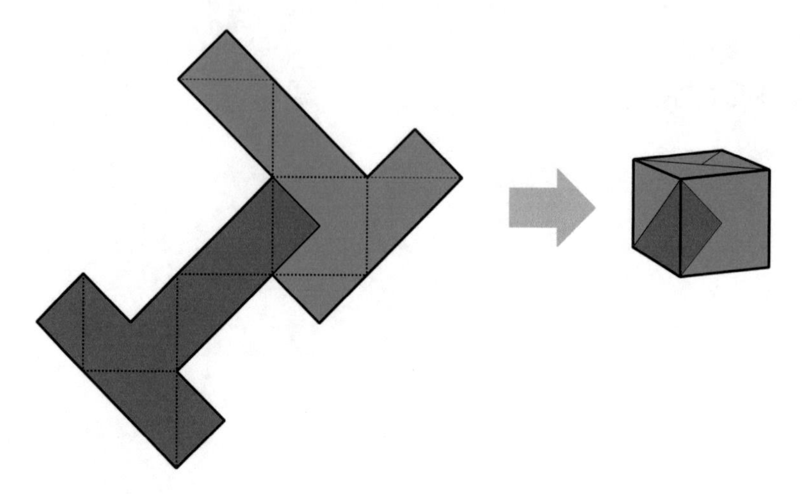

Fig. 9.1 Example of a uniform rep-cube of order 2

Now, when the nets obtained by dividing into k pieces have the same area, we say that the original net is a *regular* rep-cube of order k. Furthermore, when all the nets obtained by dividing into k pieces are congruent, the original net is called a *uniform* rep-cube of order k.

We here observe a concrete example (Fig. 9.1). These `T` shapes are one of 11 familiar nets of a cube shown in Fig. 1.1. On the other hand, if the polygon in the figure which combines these two is folded along the diagonal lines shown by the dotted line, we obtain the cube of size $\sqrt{2} \times \sqrt{2} \times \sqrt{2}$. Since the two nets are congruent, this is also an example of a uniform rep-cube. In fact, this rep-cube is also the commemorative "the first rep-cube", which I found out first at the workshop.

9.2 Regular Rep-Cubes

First we show some regular rep-cubes. Many rep-cubes are found by manual trial-and-error way (with some program search). Specifically, the following theorem is known.

Theorem 9.2.1 *For each of* $k = 2, 4, 5, 8, 9, 25, 36, 50, 64$, *there is a rep-cube of order* k.

Proof It is sufficient to show concrete examples. For small k, we show solutions in Figs. 9.1, 9.2, 9.3, and 9.4. We note that the rep-tile in Fig. 9.3 for $k = 9$ is uniform. Also note that the solution for $k = 25$, all 11 nets by edge-unfolding in Fig. 1.1 are used.

For $k = 36$, we prepare six copies of the pattern in Fig. 9.5 and arrange them in the same arrangement as one of the 11 nets of an ordinary cube. Then we can fold one

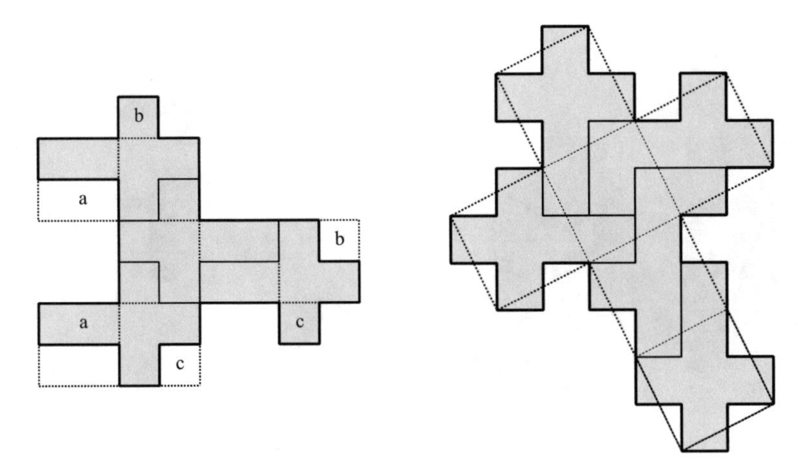

Fig. 9.2 Examples of rep-cubes of order $k = 4, 5$

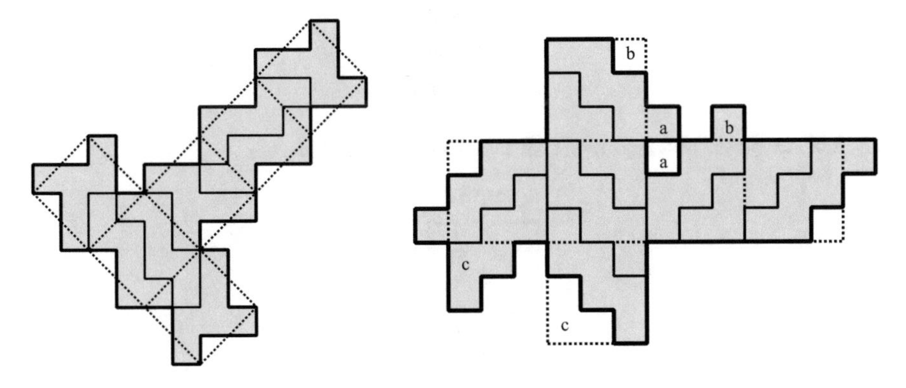

Fig. 9.3 Examples of rep-cubes of order $k = 8, 9$

cube from it which is made from 36 nets. In the case of $k = 50$, by a computer search program that pastes 50 nets to a cube of size $\sqrt{50} \times \sqrt{50} \times \sqrt{50}$ with human's help, the pattern shown in Fig. 9.7 is obtained. This pattern is also using all 11 nets by edge-unfolding of a cube. In the case of $k = 64$, we use the patterns in Fig. 9.6; using the left pattern as the bottom, pasting four copies of the center pattern as four sides, and making the right pattern as the lid, we can construct a cube of size $8 \times 8 \times 8$. By unfolding this big cube along the edges of the polyominoes, we can obtain various rep-cubes. □

Next we show that there is an infinite number of rep-cubes. First, we show that even if it is limited to uniform rep-cubes, there exist infinitely.

Theorem 9.2.2 *For any positive integer* i*, there exists a uniform regular pep-cube of order* $k = 18i^2$*. In other words, uniform and regular rep-cubes exist infinitely.*

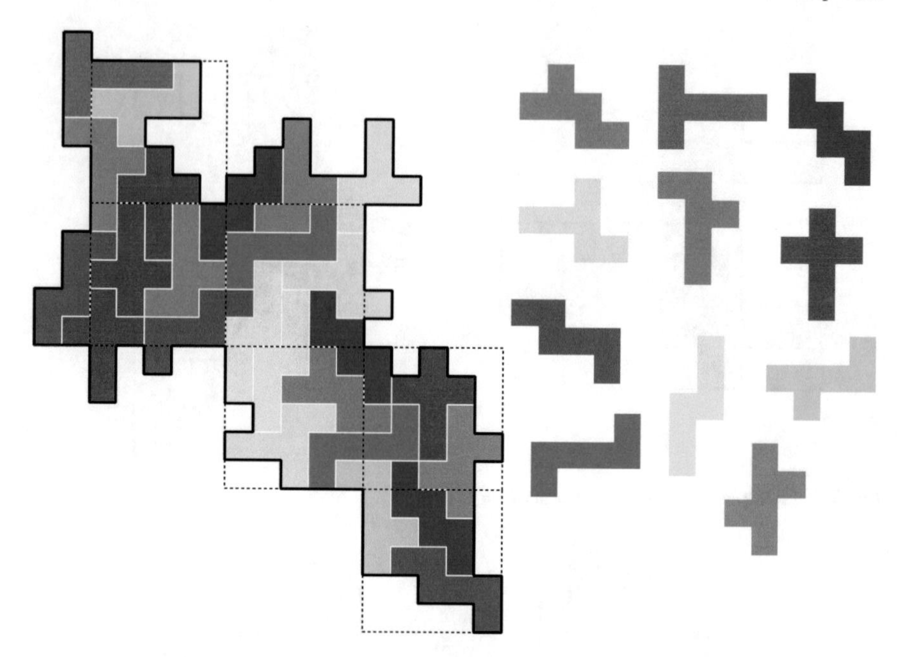

Fig. 9.4 A rep-cube of order $k = 25$. All 11 types of nets are used

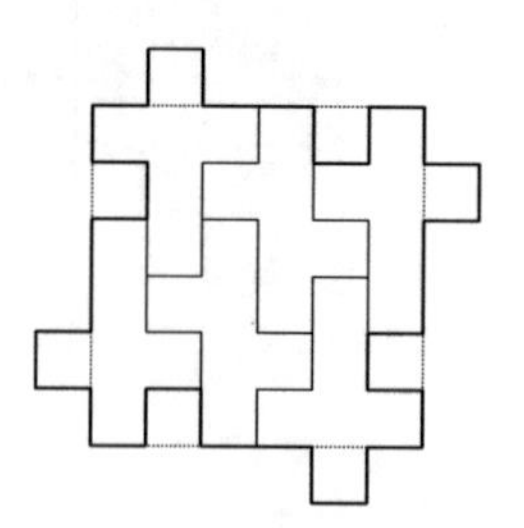

Fig. 9.5 A pattern for making a rep-cube of order $k = 36$

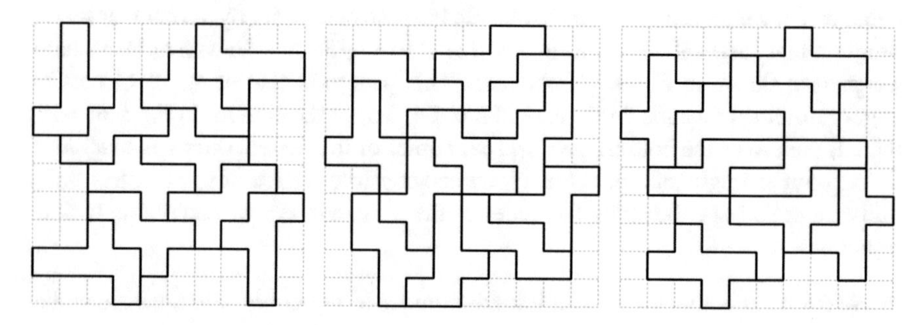

Fig. 9.6 Patterns for making a rep-cube of order $k = 64$

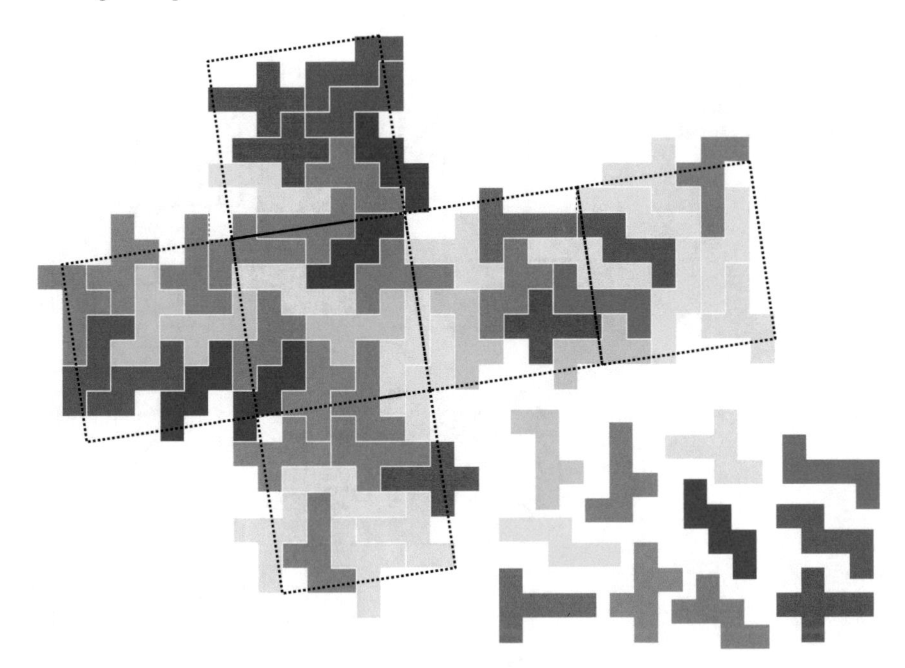

Fig. 9.7 An example of a rep-cube of order $k = 50$. All 11 types of nets are used

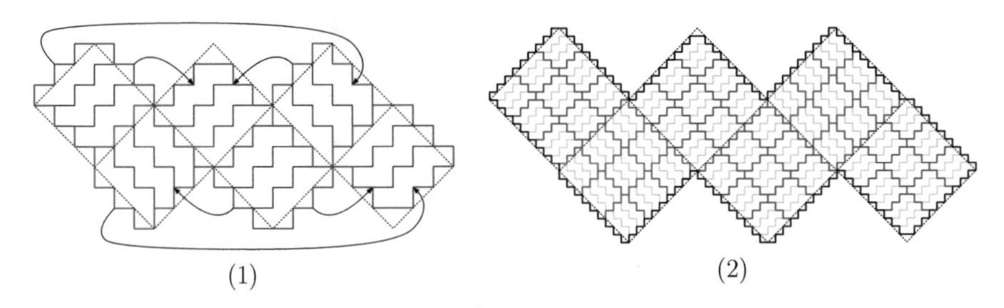

Fig. 9.8 How to construct a uniform and regular rep-cube

Proof We give a constructive proof. The construction is based on the nets used in the example of $k = 9$ in Fig. 9.3. Unfortunately, when we place the copies of it in the direction like the example of $k = 9$, it seems that we cannot tile them well. Therefore, we use the idea of placing them diagonally to the direction of the net. Then, as shown in Fig. 9.8(1), the solution for $k = 18$ is obtained. By subdividing this solution repeatedly, we can construct solutions infinitely. An example construction for $i = 3$ is shown in Fig. 9.8(2). The construction for general i is clear from the figure. □

Next, we show that the regular rep-cubes in the case of not being limited uniformly also exist infinitely. This proof is interesting because it uses a completely different

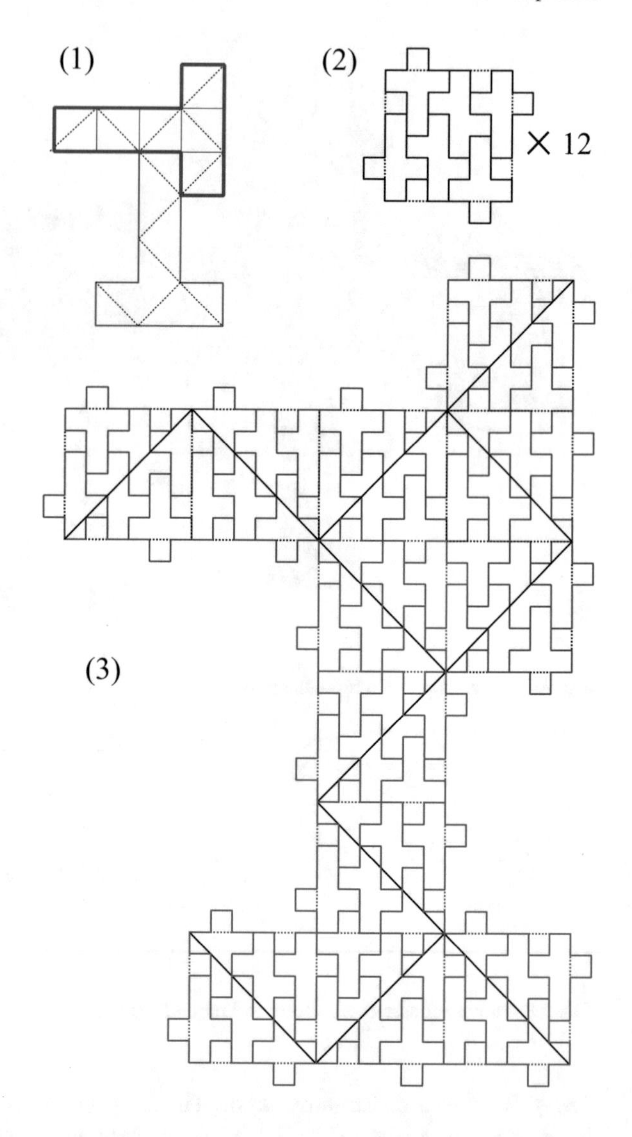

Fig. 9.9 A regular rep-cube of order 144 obtained by embedding the pattern of $k = 36$ in the pattern of $k = 2$

approach from Theorem 9.2.2. Specifically, it combines the pattern of $k = 36$ of Theorem 9.2.1 and other patterns properly.

Theorem 9.2.3 *For any integer* k', *i and any element* ℓ *of a set* $\{2, 4, 5, 8, 9, 18i^2, 25, 36, 50, 64\}$, *there is a rep-cube of order* $36\ell k'^2$. *In other words, regular rep-cubes exist infinitely.*

Proof First, according to the value of ℓ, we choose one of the patterns shown in the proofs of Theorems 9.2.1 and 9.2.2. Next, we divide the unit square in this pattern into k'^2 small squares. Furthermore, we replace each small square with the pattern for $k = 36$ in Fig. 9.5. An example is shown in Fig. 9.9. This example uses the pattern for $k = 2$ in Fig. 9.2. The zigzag lines on the boundary of this pattern do not change the property that the obtained pattern is still a net of a cube. Therefore, the resulting polygon is still a net of a cube, and the theorem holds. □

9.2.1 Enumeration of Regular Rep-Cubes

As discussed in Sect. 3.2.3, if the area is sufficiently small, we can list all polyominoes which can cover an objective box without overlapping. For example, we enumerated all common nets of two boxes of size $1 \times 1 \times 5$ and $1 \times 2 \times 3$. Using the same method, we can enumerate all regular rep-cubes with a small area. Specifically, it can be computed by Algorithm 6.

Input : An integer k;
Output: All rep-cubes of order k and area $6k$;
Let S_1 be the set of eleven edge nets of a unit cube;
for $i = 2, 3, 4, \ldots, k$ **do**
  $S_i := \emptyset$;
  for *each partial net* P *in* S_{i-1}*, do the following* **do**
    for *each edge net* P' *in* S_1*, do the following* **do**
      for *each* $\hat{P}$ *obtained by gluing an edge of* P *with an edge of* P'*, do the following* **do**
        if $\hat{P}$ *has no self-overlap and it can fold to a cube of size* $\sqrt{k} \times \sqrt{k} \times \sqrt{k}$ **then**
          Register $\hat{P}$ into S_i if it is not yet registered;
        end
      end
    end
  end
end
Output S_k;

Algorithm 6: Enumeration algorithm for rep-cubes

The algorithm checks all possible combinations in each iteration. We can save time and memory by considering symmetries. In the most part of this algorithm, we can apply the same technique shown in Sect. 3.2.3. More details can be found in [XHU17, Xu17], and omitted here.

Because we check "all possible cases" every time in the algorithm, it can actually execute only up to a small k. Table 9.1 shows the transition of the number of partial nets for the order $k = 4$. In our research group, the cases of $k = 2$ and $k = 4$ were examined. The results can be summarized as follows.

Theorem 9.2.4 *There are 33 regular rep-cubes in total with order* $k = 2$ *and area 12. There are 7185 regular rep-cubes in total with order* $k = 4$ *and area 24.*

Table 9.1 The numbers of partial nets that appear on the way of computing regular rep-cubes of area 24 with order $k = 4$

Set S_i of partial nets	S_1	S_2	S_3	S_4
Number of nets	11	2345	114852	7185

There are 33 rep-cubes in all of order $k = 2$. Among them, 17 are uniform and 16 are non-uniform. The lists are shown in Figs. 9.10 and 9.11. Among the uniform ones, four rep-cubes are rotationally symmetric.

There are 7185 regular rep-cubes of order $k = 4$, and there are 158 uniform ones in it. An example is shown in Fig. 9.12. Out of 158 uniform ones, 98 use the same net in Fig. 9.13b. The frequency in use of each of 11 nets is also shown in Fig. 9.13.

Among rep-cubes of order $k = 4$ and area 24, there are rotationally symmetric rep-cubes like ones of order $k = 2$. Moreover, there are rep-cubes having two or more dividing ways into four nets. An example is shown in Fig. 9.14. Among the data 7185, these different divisions are counted as distinct rep-cubes. However, following the original definition, they should be considered as the same rep-cubes.

When we fold a rep-cube to a cube, we can obtain a "division pattern" of the cube. Looking closely at Figs. 9.10 and 9.11, we can observe that there are different rep-cubes which are folded to the cube of the same "division pattern". When being classified by this "division pattern", there are three types of divisions that produce uniform rep-cubes, and three types of divisions that produce non-uniform rep-cubes. That is, essentially, there are six distinct types of divisions of a cube in total to produce all rep-cubes of order 2. Such a division-based approach may be advantageous to generate rep-cubes efficiently.

9.3 Cases that There Is No Regular Rep-Cube

As we have already seen, there are rep-cubes for $k = 2, 4, 5$; some can be found by hand, and it is possible to enumerate all of them for $k = 2$ and $k = 4$ by computer. Some readers may have found that $k = 3$ is missing here. In fact, when $k = 3$, there is no regular rep-cube. That is, the following theorem holds.

Theorem 9.3.1 *There is no regular rep-cube of order* 3.

Proof Let P be any polyomino (not necessarily of area 6) that can fold a cube Q. Then, by Lemma 4.2.2, P is not a convex polygon. Since the polygon P is not convex, the rolling belt (see Sect. 2.3.2) will not appear by folding P. (We do not give a proof of this fact in this book. Interested readers should refer to [DO07].) Now, when folding the cube Q from polyomino P, according to Theorem 2.1.3, the vertices of Q are on the boundary of P. Furthermore, by Theorem 2.3.5, we can assume that all vertices of Q are on grid points. Since P is a polyomino, each vertex of Q is one of the 90°, 180°, or 270° that appear at grid points on the boundary of P.

Fig. 9.10 List of regular uniform rep-cubes of order $k = 2$ and area of 12

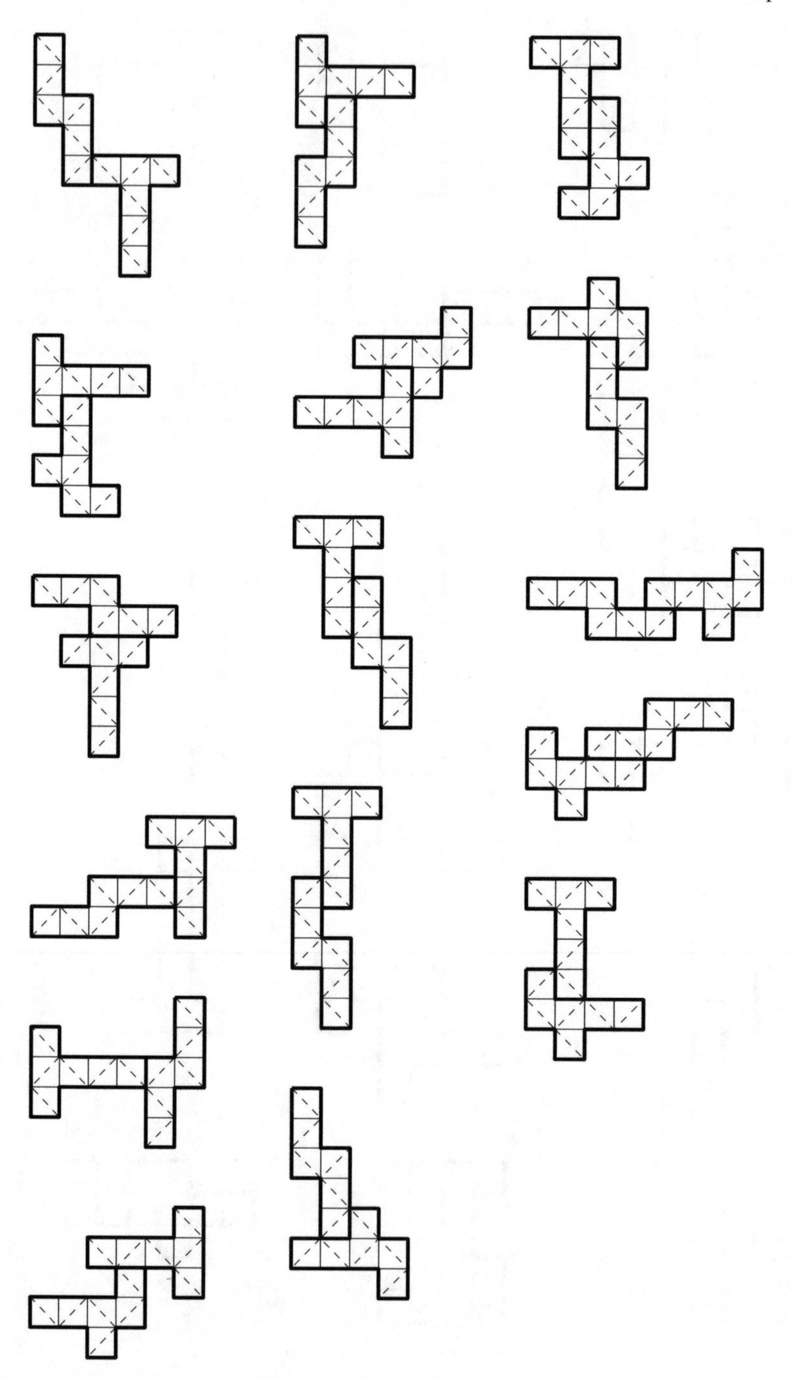

Fig. 9.11 List of regular non-uniform rep-cubes of order $k = 2$ and area of 12

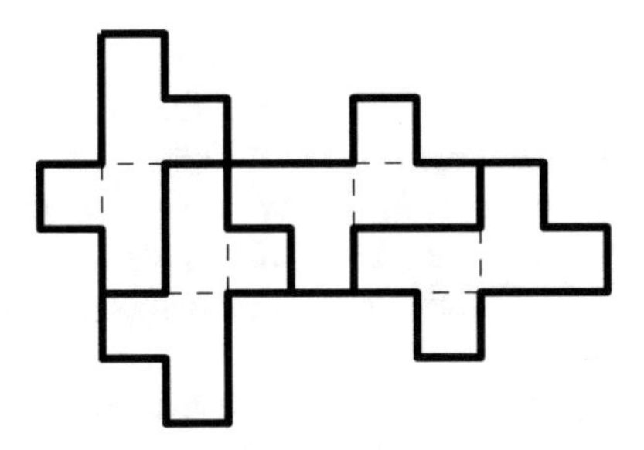

Fig. 9.12 An example of a uniform rep-cube of order $k = 4$

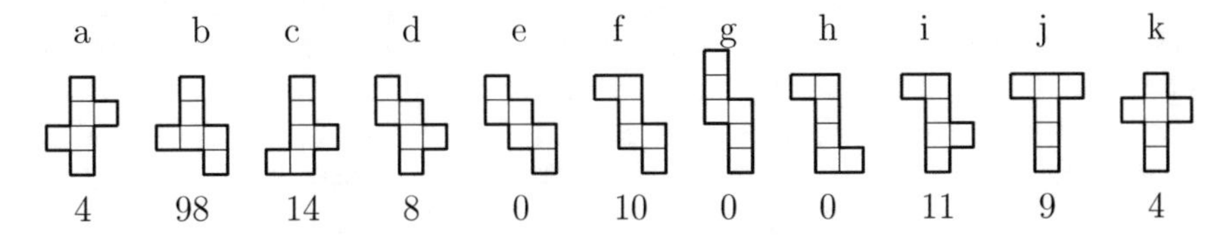

Fig. 9.13 The frequency of appearances of each net in 158 uniform rep-cubes of area 24 of order $k = 4$

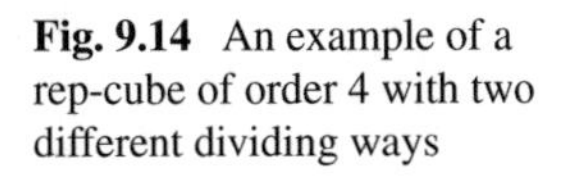

Fig. 9.14 An example of a rep-cube of order 4 with two different dividing ways

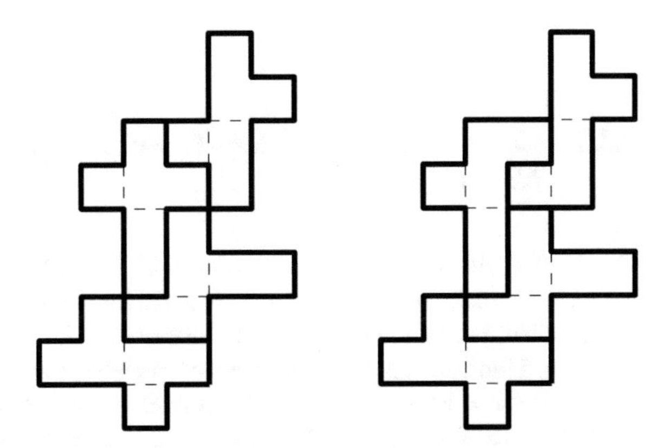

In other words, each vertex (or corner) of Q is made by either a 270° vertex alone, a 90° vertex with a 180° vertex, or three 90° vertices.

Now we move on to the proof of Theorem 9.3.1. To derive a contradiction, we assume that there is a regular rep-cube $\hat{P}$ of order 3. This polyomino $\hat{P}$ can be divided into three polyominoes P_1, P_2, P_3 of the same area, and $\hat{P}$ as well as P_1, P_2, P_3 can fold to a cube of certain size. We here let $\hat{Q}$ and Q_i be the cubes folded from $\hat{P}$ and P_i, respectively ($i = 1, 2, 3$). Let ℓ be the length of an edge of Q_i. Then P_i is a polyomino of area $6\ell^2$, and $\hat{P}$ is a palomino of area $18\ell^2$. Note that ℓ may not necessarily be an integer, but $6\ell^2$ is an integer.

Now we focus on the palomino P_1. This is a palomino of area $6\ell^2$, and we can fold it to the cube Q_1 of size $\ell \times \ell \times \ell$. By Theorem 2.3.5, we can place P_1 so that each vertex of Q_1 lies on a grid point of a square grid of size ℓ. Among the Q_1 vertices on this grid, consider two vertices v_1 and v_2 of distance ℓ. (Two vertices with a distance of ℓ may not necessarily exist, but in that case, the same argument can be made.) Then the vector $\mathbf{v_1 v_2}$ can be represented as (a, b) by the two natural numbers a and b. For these natural numbers a and b, we have $a^2 + b^2 = \ell^2$. (Similar ideas can be found in [DO07, Chap. 5.1.1] and [AHU16].) Applying the same argument to $\hat{P}$ and $\hat{Q}$, we have $\hat{a}^2 + \hat{b}^2 = 3\ell^2$ for some natural numbers $\hat{a}$ and $\hat{b}$. That is, we have $\hat{a}^2 + \hat{b}^2 = 3(a^2 + b^2)$ for them.

Therefore, it is sufficient to show that such natural numbers do not exist. To derive a contradiction, we assume that there exist natural numbers that satisfy $\hat{a}^2 + \hat{b}^2 = 3(a^2 + b^2)$. Without loss of generality, we assume that the value of this expression (that is, the value of $\hat{a}^2 + \hat{b}^2$) is the minimum positive integer that satisfies the condition. Here, $(3i \pm 1)^2 = 9i^2 \pm 6i + 1$ for any number i. Thus, for any square number x, it is a multiple of 3 or we have 1 if it is divided by 3. Here, since $\hat{a}^2 + \hat{b}^2 = 3(a^2 + b^2)$ is a multiple of 3, $\hat{a}$ and $\hat{b}$ are both multiples of 3. Thus, we can write $\hat{a} = 3\hat{a'}$ and $\hat{b} = 3\hat{b'}$ for some positive integers a' and b'. Then we obtain $\hat{a}^2 + \hat{b}^2 = (3\hat{a'})^2 + (3\hat{b'})^2 = 9(\hat{a'}^2 + \hat{b'}^2) = 3(a^2 + b^2)$, or consequently, $a^2 + b^2 = 3(\hat{a'}^2 + \hat{b'}^2)$. However, this contradicts with the minimality of the value of $\hat{a}^2 + \hat{b}^2$. Therefore, the natural numbers $a, b, \hat{a}, \hat{b}$ do not exist, and Theorem 9.3.1 is proved. □

9.3.1 Cases that Regular Rep-Cubes Are Likely to Exist and Not to Exist

Considering the argument of Theorem 9.3.1 in a slightly different viewpoint, we have a weak dichotomy of "the number that a regular rep-cube is likely to exist" and "the number that a regular rep-cube is unlikely to exist". In other words, for example, $k = 2, 4, 5$ indicate that there are regular rep-cubes, but if we clarify the property that the value of k satisfies, it is possible to predict to some extent whether rep-cubes will exist or not for a given natural number k. For example, let $\hat{P}$ be a polyomino of area $6k$ obtained by gluing k polyominoes, say $P_1, P_2, \ldots, P_k$, of area 6. In order that $\hat{P}$ is a rep-cube, natural numbers a and b such that $a^2 + b^2 = (\sqrt{k})^2 = k$ must be present. For such numbers, there is the following theorem well known in the puzzle society (see, e.g., [Ser00]).

Theorem 9.3.2 *(1) Let p be a prime number. Then the necessary and sufficient condition that can be represented as $p = a^2 + b^2$ by two non-negative integers a, b is either $p = 2$ (when $a = b = 1$) or $p \equiv 1 \pmod 4$. (2) Let x be a composite number and the result of factoring of x is $p_1^{d_1} p_2^{d_2} \cdots p_m^{d_m}$ for some prime numbers $p_1, \ldots, p_m$. Then the necessary and sufficient condition that x can be represented as $x = a^2 + b^2$ by two non-negative integers a, b is that d_i is an even number for each prime factor p_i that satisfies $p_i \equiv 3 \pmod 4$.*

Theorem 9.3.2(1) is known as "Fermat's theorem on sums of two squares", which was first proposed as a problem by Fermat, and Euler gave the proof.

Although it can be determined by Theorem 9.3.2, it seems to be useful to list the numbers here that regular rep-cubes can exist. Specifically, if we calculate $0 \leq a \leq b$ from the smaller ones, we have $k = 2$, 4, 5, 8, 9, 10, 13, 16, 17, 18, 20, 25, 26, 29, 32, 34, 36, 37, 40, 41, 45, 49, 50, 52, 53, 58, 61, 64, 65, 68, 72, 73, 74, 80, 81, 82, 85, 89, 90, 97, 98, 100 . . ., and so on. It seems that there is a regular rep-cube for these numbers, and we can observe that this sequence includes the numbers $2, 4, 5, 8, 9, 25, 36, 50, 64$ that appear in Theorem 9.2.1. Conversely, numbers that do not appear here, such as $k = 3, 6, 7, 11, 12, 14, 15, 19, 21, 22, 23$, are considered to have no regular rep-cube. We proved specifically in the case of $k = 3$, but the general method to handle these numbers uniformly is unsolved.

Open Problem 9.3.1 *For a given integer k, is there an efficient way to determine if there is a regular rep-cube of order k? It may be easier to prove that there is no regular rep-cube for the numbers that do not meet Theorem 9.3.2 (such as* $k = 6, 7, 11, 12$*).*

9.4 Extensions to Non-Regular Rep-Cubes and Pythagorean Triples

First, two non-regular rep-cubes are constructed by trial and error.

Theorem 9.4.1 *There are non-regular rep-cubes of order* $k = 2$, 10.

Proof Each rep-cube is shown in Fig. 9.15.

The non-regular rep-cube of order $k = 2$ is the left one in Fig. 9.15. As a whole, a cube of size $\sqrt{5} \times \sqrt{5} \times \sqrt{5}$ can be folded; on the other hand, if it is divided into two, one can be folded to a cube of size $2 \times 2 \times 2$, and the other one can be folded to a unit cube. The relationship of the areas is $6 \times (\sqrt{5})^2 = 6 \times 1^2 + 6 \times 2^2 = 30$.

The other non-regular rep-cube of order $k = 10$ is the right one in Fig. 9.15. This pattern has an area of 150. It is easy to see the nets of the nine unit cubes. Their areas occupy 54 squares in total. The remaining 96 squares make up a net of a cube of size $4 \times 4 \times 4$. As a whole, it is a net of a cube of size $5 \times 5 \times 5$. These areas satisfy the relation of $150 = 6 \times 5^2 = 6 \times (3^2) + 6 \times (4^2) = 6(3^2 + 4^2)$. □

9.4.1 *How to Construct Non-Regular Rep-Cubes*

Next, we show a constructive way to generate non-regular rep-cubes.

Theorem 9.4.2 *There is an infinite number of non-regular rep-cubes. Specifically, when there is a regular rep-cube of order k, we can construct a non-regular rep-cube of order* $k + 35i$ *for any positive integer* i.

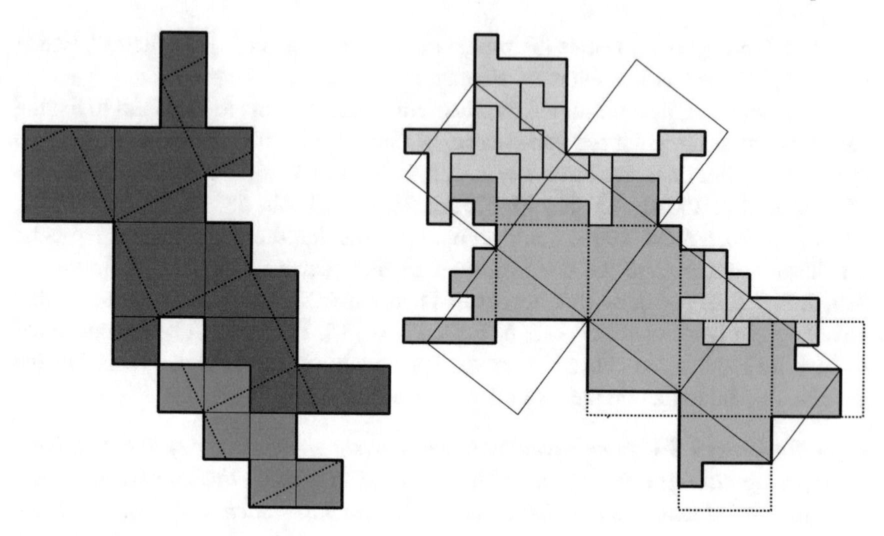

Fig. 9.15 Non-regular rep-cubes of order $k = 2, 10$

Proof This construction is essentially an application of Theorem 9.2.3. In Theorem 9.2.3, all the unit squares in a net are replaced at once with the pattern shown in Fig. 9.5. However, we do not need to apply this replacement to all of them. Let us explain by giving a more specific example. Figure 9.16(1) is the rep-cube of order 2 shown in Fig. 9.1. We replace only the squares that appear in the above net with the pattern shown in Fig. 9.5. Then this net is split into 36 small nets. At that time, the boundary of the original net becomes a zigzag line from a straight line. Therefore, we modify the boundary of the other net to fit it. Figure 9.16(2) obtained in this way is a non-regular rep-cube of order 37. From the smallest net of these 37 nets, if one is appropriately selected and the same procedure is carried out, another non-regular rep-cube of order 72 shown in Fig. 9.16(3) is obtained. Repeating this process, we can generate infinitely non-regular rep-cubes. □

The above construction method has a very high degree of freedom, and it seems to be possible to generate in other different methods. From this intuition, the following open problem arises.

Open Problem 9.4.1 *For sufficiently large k, there may be a non-regular rep-cube of order k. In other words, does there exist an integer K such that there is a rep-cube of order k for any integer $k > K$?*

There is little else known about non-regular rep-cubes; however, research is currently underway, with the expectation that it is likely to be particularly relevant to the Pythagorean triple. We here introduce what we know about the relationship with the Pythagorean triple.

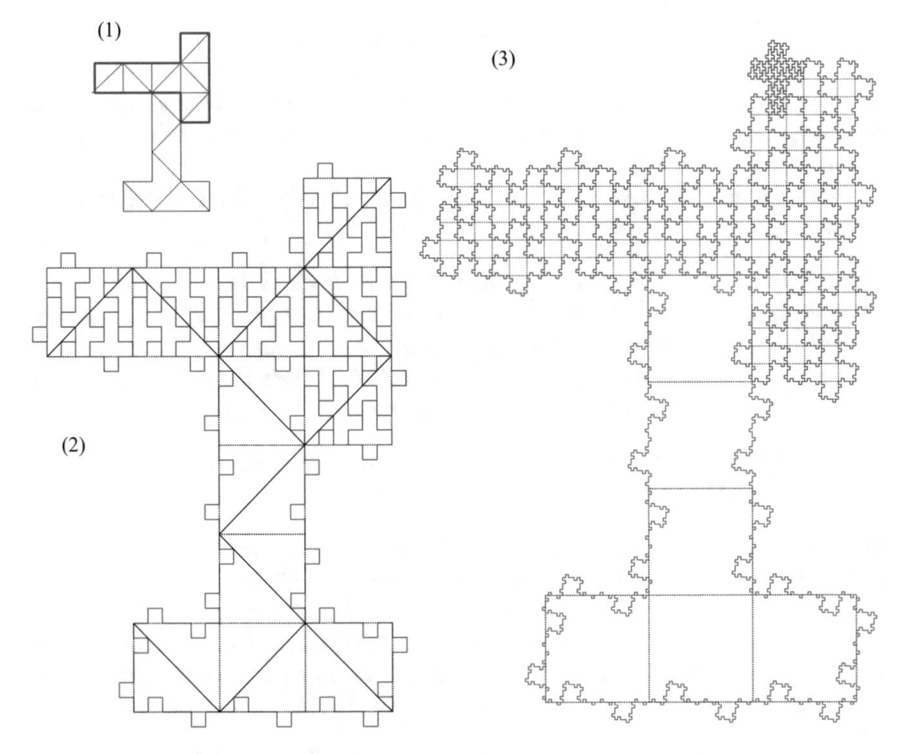

Fig. 9.16 How to generate an infinite of non-regular rep-cubes

9.4.2 Pythagorean Triples

First, let us introduce *Pythagorean triple*. The Pythagorean triple is a triple (a, b, c) of natural numbers that satisfy $a^2 + b^2 = c^2$. For example, (3, 4, 5) and (5, 12, 13) are representative Pythagorean triples since we have $3^2 + 4^2 = 9 + 16 = 25 = 5^2$ and $5^2 + 12^2 = 25 + 144 = 169 = 13^2$, respectively. Hereafter, we assume $a < b < c$ without loss of generality. At first glance, it seems to be hard to find such a triple, but in fact the following characterization has been known for the Pythagorean triple from old times.

Theorem 9.4.3 *Suppose two positive integers m and n satisfy the following three conditions: (1) m and n are relatively prime, (2) $0 < n < m$, and (3) $m - n$ is odd. Then $(m^2 - n^2, 2mn, m^2 + n^2)$ forms a Pythagorean triple. Conversely, any Pythagorean triple (a, b, c) can be represented in this form for some m and n which satisfy these three conditions.*

The proof of Theorem 9.4.3 is elementary, so we omit it. If we find m and n in order in the three conditions, we can generate any number of Pythagorean triples. For example, letting $m = 2$ and $n = 1$, we have (3, 4, 5), and $m = 3$ and $n = 2$ give us

(5, 12, 13). Note that $m^2 + n^2$ is the largest among the triples $(m^2 - n^2, 2mn, m^2 + n^2)$, so we have $c = m^2 + n^2$. However, depending on how we choose m and n, which of $m^2 - n^2$ and $2mn$ is larger will change. Thus, when we consider a Pythagorean triple (a, b, c) with $a < b < c$, it is different in some cases whether $m^2 - n^2$ or $2mn$ will be a and which will be b.

Now, for any Pythagorean triple (a, b, c), we trivially have $6a^2 + 6b^2 = 6c^2$, which leads us to the following natural problem.

Open Problem 9.4.2 *For a Pythagorean triple (a, b, c), is there a corresponding rep-cube of order 2? In other words, is there a polyomino P_c of area $6c^2$ that satisfies the following two conditions? (1) A cube of size $c \times c \times c$ can be folded from P_c. (2) P_c can be divided into two polyominoes P_a and P_b such that a cube of size $a \times a \times a$ can be folded from P_a, and a cube of size $b \times b \times b$ can be folded from P_b.*

This problem is an unsolved problem. Then we relax the above condition (2) a little and consider the following minimization problem.

Open Problem 9.4.3 *For a Pythagorean triple (a, b, c), consider a polyomino P_c of area $6c^2$ that satisfies the following two conditions. (1) A cube of size $c \times c \times c$ can be folded from P_c. (2) P_c is divided into k polyominoes $P_1, \ldots, P_k$ such that some of them are glued to make a polyomino P_a which can be folded to a cube of size $a \times a \times a$, and a cube of size $b \times b \times b$ can be folded from a polyomino P_b made by gluing the rest. For a given Pythagorean triple, find the minimum k.*

This is immediately obvious that there exists a solution; dividing the polyomino P_c into $6c^2$ unit squares, we have a trivial upper bound $k = 6c^2$. In the original open problem, $k = 2$ is the goal, and this has been re-formulated as a minimization problem. Then how much k can be reduced from the obvious solution $6c^2$. So far $k = 5$ is the best achievement. We introduce this impressive current record.

9.4.3 *Five-Piece Solution for Pythagorean Triple*

The main theorem is as follows.

Theorem 9.4.4 *Let (a, b, c) of integers be any Pythagorean triple that satisfies $a < b < c$. Then, there is a set $S(a, b, c)$ of five polyominoes that satisfy the following condition. (1) A cube of size $c \times c \times c$ can be folded when these five polyominoes are put together. (2) From one of them, a cube of size $a \times a \times a$ can be folded, and a cube of size $b \times b \times b$ can be folded if the remaining four polyominoes are put together.*

That is, Theorem 9.4.4 gives the solution $k = 5$ to the optimization Problem 9.4.3. We note that Theorem 9.4.4 holds for general Pythagorean triples. Let us give an example in Fig. 9.17 to get a sense. Here is the solution for the Pythagorean triple (3, 4, 5). Figure 9.17a is a polyomino that can fold to a cube of size $3 \times 3 \times$

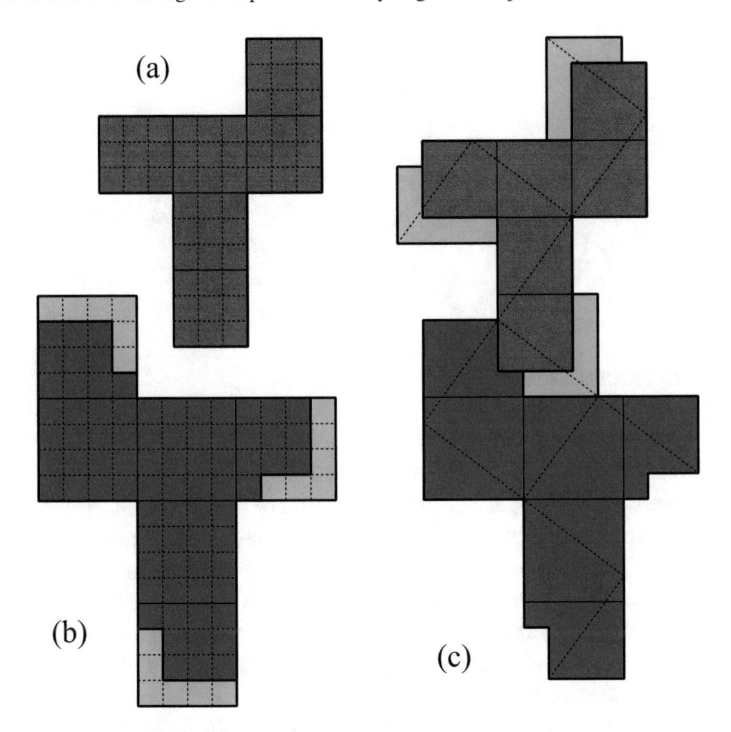

Fig. 9.17 Five-piece solution for the Pythagorean triple (3, 4, 5). **a** A net of a cube of size $3 \times 3 \times 3$. **b** A net of a cube of size $4 \times 4 \times 4$. **c** A net of a cube of size $5 \times 5 \times 5$

3, and Fig. 9.17b is a polyomino that can fold to a cube of size $4 \times 4 \times 4$. The polyomino in Fig. 9.17c is a polyomino by gluing the polyomino in Fig. 9.17a with four polyominoes made by cutting one in Fig. 9.17b. Although this may not be visible so much, if you draw actually on paper, cut out, and fold along a dotted line, you will immediately understand that the polyomino in Fig. 9.17c can be folded to a cube of size $5 \times 5 \times 5$. We will show that such a construction is generally possible for *any* Pythagorean triple.

Proof First, the rough idea of the construction method is depicted in Fig. 9.18. The first step is to open a cube of size $a \times a \times a$ and a cube of size $b \times b \times b$ symmetrically from one vertex. Three edges are connected to one vertex, so we cut them up to the adjacent vertex. We cut three more edges further from these three vertices adjacent to the first vertex, but we have to cut them in the same direction, otherwise the cuts will collide with each other. Thus, align the directions and cut up to three vertices further away, so that the cuts do not hit each other. Then expand the cube along the line we have cut so far. Three squares are opened from the first vertex that starts cutting, and each square is connected separately to the other three squares that are gathered at the opposite vertex. At this point, the solid is an open tetra cone shape made by three 1×2 rectangles connected like a windmill. This windmill shape is rotationally symmetrical around the vertex where three rectangles are gathered.

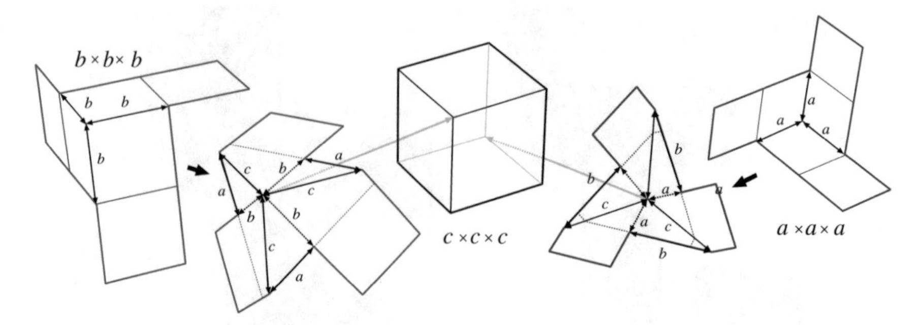

Fig. 9.18 A rough idea of construction

Then, if we spread the paper around well, it will be shaped like a cone with this vertex as the center or the apex. The cone's boundary forms a bump, as we see at the other side of three rectangles. For convenience, we name these bumpy cones S_a and S_b made from two cubes of size $a \times a \times a$ and $b \times b \times b$, respectively. The paper around the apex of this bumpy cone has a total of 270°, so if we make appropriate creases, the apex (or the corner of the cube) will fit perfectly to a corner of the cube of size $c \times c \times c$. Now, we consider mapping the apices of S_a and S_b exactly on the two opposite vertices of a cube of size $c \times c \times c$.

The most important point of this construction is to *twist* when pressing the apices of S_a and S_b to the vertices of the cube of size $c \times c \times c$. That is, the point is that the square grids of the original two small cubes do **not fit** to the square grid on the large cube. When we twist them properly, we can place all the vertices of a large cube on the boundaries of S_a and S_b. The crucial point of this construction is to create a Pythagorean triangle whose length of three edges is (a, b, c) by making different crease lines on the twisted and pasted cones. There are two possible ways of construction depending on the size of the two small cubes.

Case $a < b < 2a$: For example, the most famous Pythagorean triple $(3, 4, 5)$ is in this case. This case illustrated on the net of a cube of size $c \times c \times c$ is shown in Fig. 9.19. The entire outer frame is a net of a cube of size $c \times c \times c$. Here, the three vertices labeled p gather at one vertex on the large cube, and the apex of a cubed cone S_a is attached to the point, where the cone S_a is obtained by opening the small cube of size $a \times a \times a$. In the figure, the six faces of this small cube are already drawn, and the boundary of this S_a forms a zigzag line connecting two points X_1 and X_2. The three opposite vertices labeled q gather at the other vertex of the large cube. Here is the point where the apex of a cubed cone S_b is attached, and S_b is obtained by opening the middle cube of size $b \times b \times b$. In the figure, only three middle square faces are drawn. A zigzag line formed by these three squares connects two points Y_1 and Y_2. Therefore, all we have to do here is to create three more squares of size $b \times b$, with as few splits as possible, from the belt part sandwiched between the zigzag lines X_1X_2 and Y_1Y_2.

First, as shown in Fig. 9.19, we construct the grid by extending the edges of a square of size $b \times b$. Then, the belt part sandwiched by the zigzag lines is divided

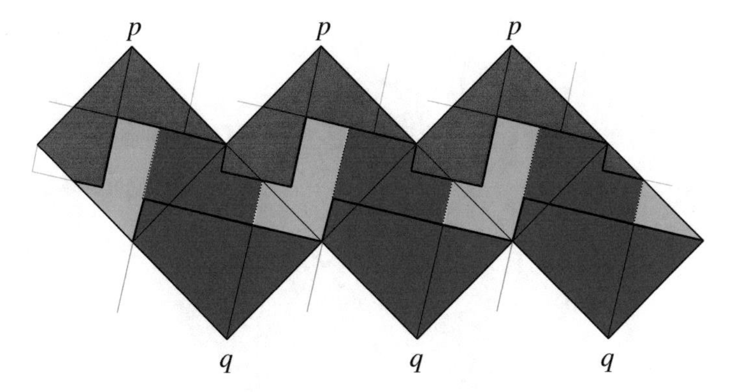

Fig. 9.19 A net of a cube of size $c \times c \times c$

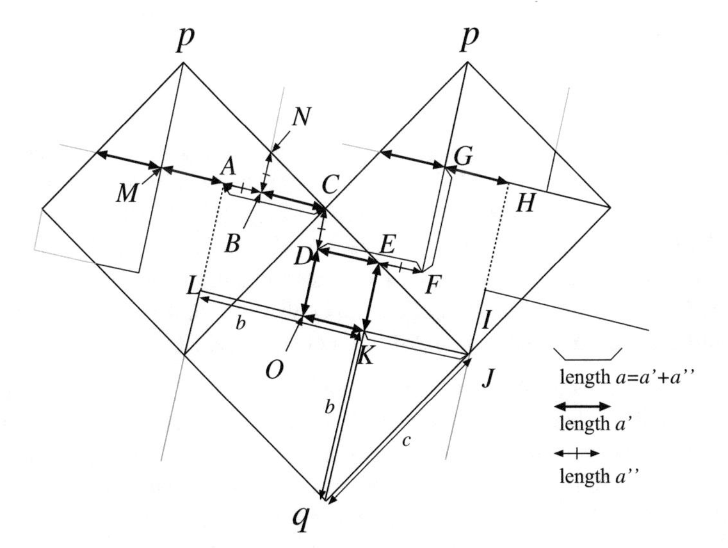

Fig. 9.20 Details of the lengths in polyomino

into six parts. Specifically, there is a hexagon $ACDEKL$ with two more congruent ones, and a hexagon $EFGHJK$ with two more congruent ones. What is shown here is that when gluing two hexagons $ACDEKL$ and $EFGHJK$ at line segments $GFEK$ and $ACDE$, we always obtain a square $HJKL$ of size $b \times b$. If this claim is correct, clearly, the theorem immediately follows.

Now we look at the corresponding part of the belt in detail (Fig. 9.20). First, let xyz be a right triangle with three edges of length $|xy| = a$, $|yz| = b$, and $|zx| = c$. Then the two triangles pMC and JKq of the belt are congruent with this triangle xyz. For simplicity, we write $a' = b - a$ and $a'' = a - a' = 2a - b$. Then since $|MC| = b$ and $|MB| = a$, we have $|BC| = b - a = a'$. When we fold a small cube, the edges BC and CD are forming two edges of square of size $a \times a$; therefore, we

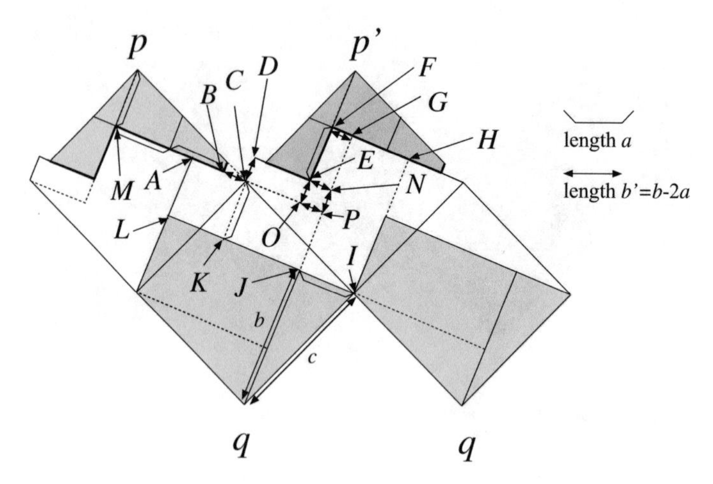

Fig. 9.21 Shapes when $2a < b$

have $|CD| = a - a' = a''$. Since triangle NBC and triangle CDE are congruent, $|DE| = a'$ and thus $|EF| = a''$. Since triangle COJ is congruent with the right triangle xyz, $|CO| = a$ and $|DO| = a'$, so $|EK| = a'$. Here $|EF| = a''$ and $|KJ| = a$ yield $|GH| = |MA| = a'$. Thus, $|AC| = b - a' = a$. Therefore, the zigzag lines $ACDE$ and $GFEK$ can be glued exactly, as their corresponding lengths are all the same and their angles are at right angles. Given that the length of the side $|LK|$ is b and the area of the entire belt, the resulting rectangle $LKJH$ is a square.

Case $2a < b$: For example, the Pythagorean triple (5, 12, 13), which is next to (3, 4, 5), satisfies this condition. The basic idea of proof is the same as the case $a < b < 2a$. However, in the present case, the shape of the belt changes because $a'' = 2a - b < 0$ in Fig. 9.20. Anyway, we first glue six squares of size $a \times a$ and three squares of size $b \times b$ on the net of a large cube of size $c \times c \times c$. A part of the state of the net at this time is shown in Fig. 9.21 (the glued squares are represented in gray). In the figure, to improve visibility, the cut lines around small squares of size $a \times a$ are changed to zigzag lines. In fact, note that this argument can be proceeded on the cube of size $c \times c \times c$ without unfolding. Since the shape of the belt is slightly different from that of the case $a < b < 2a$, it is necessary to change the way of drawing the cut lines according to the shape, but it is the same thing. Namely, we construct three more squares of size $b \times b$ from this belt part with a small number of pieces. In this case, we cut along the line segment ENJ in the figure to make a square of size $b \times b$.

First, we confirm that two triangles qIJ and ICK are congruent to the right triangle xyz, where $|xy| = c$, $|yz| = a$, and $|zx| = b$. We also confirm that JL is one edge of a square of size $b \times b$. Therefore, if we pay attention to $|IK| = |JL| = b$ and $|IJ| = a$, we obtain $|IJ| = |KL| = a$. Thus $|AC| = a$. We let b' be the length of $|CD|$. Then, since the edge CD is glued to the edge BC and that the triangle CpM is congruent to the right triangle xyz, we have $|CD| = |BC| = b' = b - 2a$.

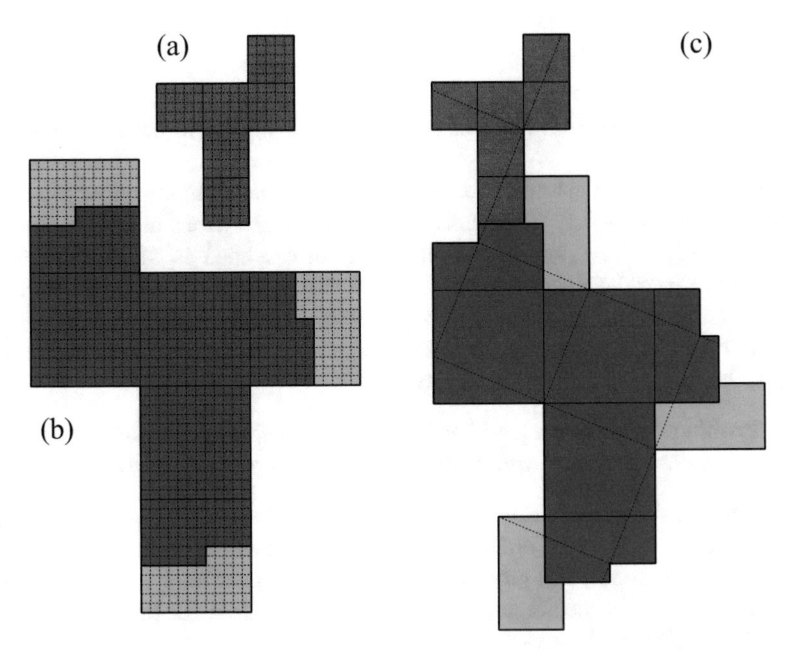

Fig. 9.22 Five-piece solution for the Pythagorean triple (5, 12, 13). **a** A net of a cube of size $5 \times 5 \times 5$. **b** A net of a cube of size $12 \times 12 \times 12$. **c** A net of a cube of size $13 \times 13 \times 13$

On the other hand, since the triangle $p'CO$ is congruent to right triangle xyz, $|DE| = |CO| = a$. Focusing on a square of size $a \times a$, we get $|EF| = a$. Furthermore, $|FG| = |IK| - |CO| - |GH| = b - 2a = b'$. Thus, the rectangle $ENPO$ is actually a square of size $b' \times b'$ and $|DN| = a + b' = b - a$. From these facts, we have $|AC| = |EF| = a$, $|CD| = |EN| = b' = b - 2a$, and $|DN| = |NJ| = a + b' = b - a$. Hence, if we glue the zigzag line $ACDN$ of the hexagon $ACDNJL$ to the zigzag line $FENJ$ of the hexagon $FHIJNE$, we get a square $HIJL$ of size $b \times b$.

For example, from the case of Pythagorean triple (5, 12, 13), we can obtain the polyominoes shown in Fig. 9.22.

From above analysis, we obtain the theorem in both cases. We note that both $a = b$ and $2a = b$ are not feasible by the constraints on n and m. □

Theorem 9.4.4 leads us to the following corollary.

Corollary 9.4.5 *There are infinitely many sets of five polyominoes and triples (a, b, c) of positive integers that satisfy the following conditions: (1) By gluing these five polyominoes, we obtain a net of a cube of size $c \times c \times c$. (2) One polyomino is a net of a cube of size $a \times a \times a$, and we obtain a net of a cube of size $b \times b \times b$ by gluing the remaining four polyominoes.*

9.5 Open Problems

As for rep-cubes, research has just begun in 2016, and there are many unsolved problems.

First problem is the classification of integers with and without regular rep-cubes. As we saw in Sect. 9.3.1, "the numbers that regular rep-cubes are likely to exist" are $k = 2, 4, 5, 8, 9, \ldots$, and this sequence can be extended as we like. Of course, this sequence includes the integers 2, 4, 5, 8, 9, 36, 50, 64 that appeared in Theorem 9.2.1. So far, however, we do not know how to generate a regular rep-cube of order k for these numbers, in general, except finding a real polyomino for each specific k. That is, the following problem is unsolved.

Open Problem 9.5.1 *For a positive integer k that satisfies* $a^2 + b^2 = (\sqrt{k})^2 = k$ *for some positive integers a and b, does a regular rep-cube of order k always exist?*

Conversely, for integers excluded from the above sequence, can we always say that there is no rep-cube of order k? Intuitively, it seems impossible to construct a rep-cube in such a case. The proof of Theorem 9.3.1 argues for $k = 3$, and we use a few properties of this specific number 3. Can we generalize this proof? That is, the following problem is also unsolved.

Open Problem 9.5.2 *For a positive integer k such that there are no positive integers a and b with* $a^2 + b^2 = (\sqrt{k})^2 = k$*, prove that there is no regular rep-cube of order k.*

Through a series of studies, some special regular rep-cubes are found, which are "uniform rep-cubes" and "rotationally symmetric rep-cubes". For uniform rep-cubes, the following problems are interesting.

Open Problem 9.5.3 *There are 11 representative nets of a cube by edge-unfolding (Fig. 1.1). Is there a uniform rep-cube for each of these 11 nets?*

Uniform rep-cubes for 9 of 11 edge nets have already been found (one in Fig. 9.8 and eight in Fig. 9.13), but what about remaining two (in Fig. 9.13g and h)?

On the contrary, what about rep-cubes that use all 11 nets by edge-unfolding? Since the rep-cube of order 25 shown in Fig. 9.4 and the rep-cube of order 50 shown in Fig. 9.7 contain all of them, there is a solution to this problem. That $k = 25$ is the smallest solution up to now, but can we make k smaller? Mr. Jun Maekawa, who is famous as an origami designer, suggested the following puzzle to me: Let us distinguish the back and the front of 11 nets of a cube. Then, there are two different shapes of back and front for nine types, except the `T` shape and the Latin cross shape. Collecting all these shapes gives $9 \times 2 + 2 = 20$ nets. That is, there are 20 different shapes of nets if we distinguish back and front. This area totals 120, which gives us the length of an edge equal to $2\sqrt{5}$; thus, as we considered the diagonal length of $\sqrt{5}$ in Sect. 3.4.1, when we fold along diagonal lines, they might be folded to a cube. Does such a rep-cube exist?

Open Problem 9.5.4 *There are 11 representative nets of a cube. When* $k = 25$ *or* $k = 50$*, there is a rep-cube of order* k *that contains all types. What is the minimum of such* k*? In particular, is there a rep-cube of order* $k = 20$ *that meets Maekawa's requirement?*

We do not know much about non-regular rep-cubes.

Even for the order k in which there may be a non-regular rep-cube, the methods of combinations increase explosively because the number of division is diverse. For example, considering only the areas, a net of a cube of size $c \times c \times c$ may be divided to a net of a cube of size $b \times b \times b$ and $c^2 - b^2$ nets of a unit cube. However, is such a rep-cube always exist? For example, the following problems can be considered.

Open Problem 9.5.5 *There are an infinite number of non-regular rep-cubes; however, as a special example, what about the following two? (1) Does a rep-cube corresponding to the Pythagorean triple* (a, b, c) *always exist? (2) For positive integers* $1 < b < c$*, is there a rep-cube which can be folded to a cube of size* $c \times c \times c$ *and can be divided into one net of a cube of size* $b \times b \times b$ *and* $c^2 - b^2$ *nets of a unit cube?*

Here, looking back at the definition of rep-cube, a rep-cube is a polyomino P such that (1) P itself is a net of a cube, and (2) when we divide P into k polyominoes $P_1, P_2, \ldots, P_k$, they were each a net of a cube. However, observing the solutions of enumeration of regular rep-cubes of order $k = 4$ in Sect. 9.2.1, we found a net P that has two ways to divide it into four nets (Fig. 9.14). In this case, the meaning of a polyomino P is a rep-cube becomes vague. We may say that this polyomino P contains *two* rep-cubes. To avoid this ambiguity, it may be necessary to reconsider the rep-cube as "a division pattern of the surface of a cube". This seems to be a future research topic rather than an unsolved problem.

9.6 Extension to Doubly Covered Square and Regular Tetrahedron

When we naturally extend the idea of dividing a net into nets, we can consider other problems. Here we take doubly covered squares and regular tetrahedra as natural extensions. Here, a *doubly covered square* is a figure obtained by gluing four corresponding edges of the two squares together. Although there may be a sense of discomfort, it is considered as a basic "polyhedron" in the field of mathematics. On a doubly covered square made by two unit squares, the surface area is 2 and the volume is 0.

Before dealing with these two polyhedra, we show a basic lemma that can be used in common.

Lemma 9.6.1 *Let* P *be a cylinder whose circumference is* a *and whose height is* b*. (It only has one side without lid and bottom.) For any angle* $0 < \theta \leq 90°$*, we define*

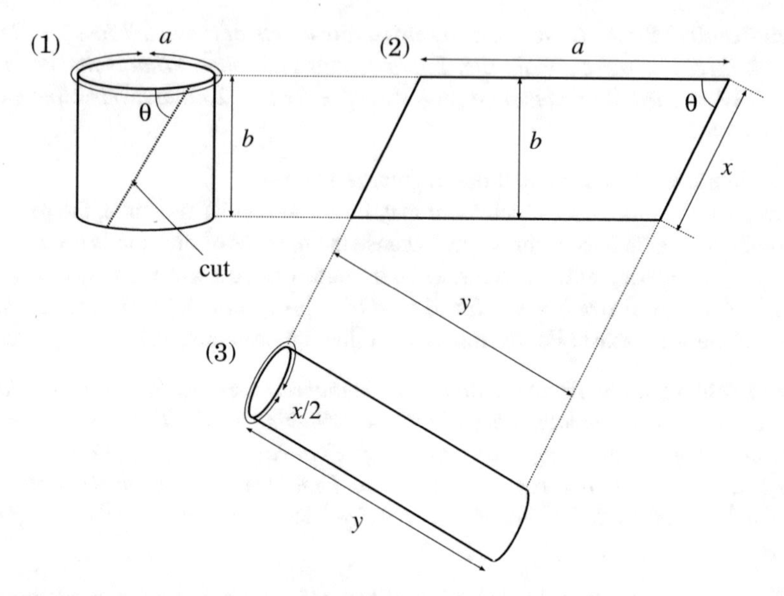

Fig. 9.23 (1) A cylinder of circumference a height b. (2) Common net of two cylinders. (3) Another cylinder of circumference $x/2$ and height y

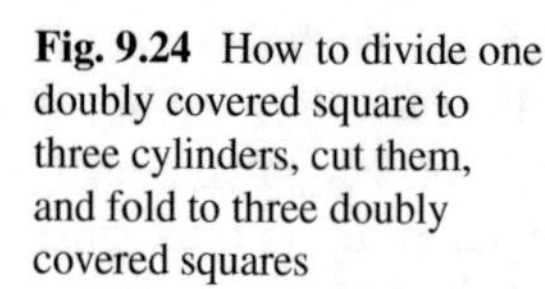

Fig. 9.24 How to divide one doubly covered square to three cylinders, cut them, and fold to three doubly covered squares

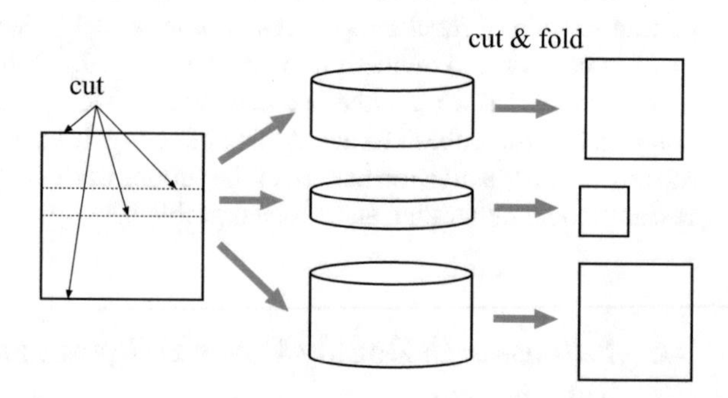

$x = \frac{b}{\sin\theta}$ and $y = a\sin\theta$, and let Q be a cylinder of circumference $x/2$ and height y. Then P and Q have a common net.

Proof The construction of the cylinder Q from P is shown in Fig. 9.23. First, we cut P along the dotted line in Fig. 9.23(1). This gives us a parallel quadrilateral with sides a and $x = \frac{b}{\sin\theta}$. Rewinding along the edge of length x, a cylinder Q of the desired size can be obtained as shown in Fig. 9.23(3). □

Note that if the angle θ is as close as 90° to 0°, x can be set as large as $b/2$ or more.

Now we turn to the discussion of doubly covered squares. Using the above lemma, we can prove the following theorem for doubly covered squares.

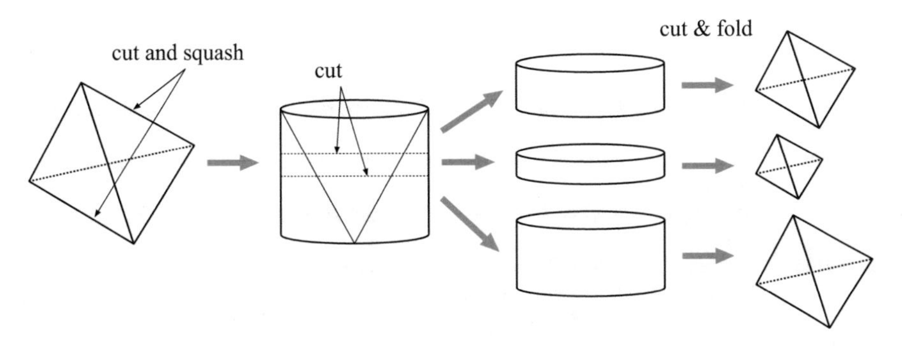

Fig. 9.25 How to divide one regular tetrahedron to three cylinders, cut them, and fold to three regular tetrahedra

Theorem 9.6.1 *For any sequence of positive real numbers $a_1, a_2, \ldots, a_k$ with $\sum_i a_i = A$, there is a net P of a doubly covered square of area A that satisfies the following conditions: (1) P can be divided into k polygons of area $a_1, a_2, \ldots, a_k$ and (2) each polygon can fold to a doubly covered square.*

Proof First, we divide the doubly covered square of area A into k pieces with $k - 1$ horizontal lines parallel to the lower side so that each piece has area $a_1, a_2, \ldots, a_k$, respectively (Fig. 9.24). As a result of this division, two envelope shapes (one of two edges of a cylinder is glued) corresponding to areas a_1 and a_k, and $k - 2$ cylinders of areas $a_2, \ldots, a_{k-1}$ are obtained. Cut one side of these two envelope shapes into a cylinder to adjust the others (Fig. 9.24).

Now we focus on the ith cylinder of area a_i. The circumference of this cylinder is $2(\sqrt{A/2}) = \sqrt{2A}$, so its height is $a_k/\sqrt{2A}$. Since $a_k < A$, we have $a_k/\sqrt{2A} < \sqrt{a_k/2}$. Therefore, by Lemma 9.6.1, this ith cylinder and a cylinder of circumference $2\sqrt{a_k/2}$ and of height $\sqrt{a_k/2}$ have a common net. Constructing this common net for each i and gluing all of them side by side, we obtain the desired P. □

The method used in Theorem 9.6.1 can be applied to tetrahedra as it is.

Theorem 9.6.2 *For any sequence of positive real numbers $a_1, a_2, \ldots, a_k$ with $\sum_i a_i = A$, there is a net P of a regular tetrahedron of area A that satisfies the following conditions: (1) P can be divided into k polygons of area $a_1, a_2, \ldots, a_k$ and (2) each polygon can fold to a regular tetrahedron.*

Proof In the same way as Fig. 2.11, a cylinder is obtained by cutting two edges of the tetrahedron in the positions of the opposite sides. When calculating the size of this cylinder, the circumference is $\sqrt{\frac{4A}{\sqrt{3}}}$ and the height is $\sqrt{\frac{\sqrt{3}A}{4}}$. By applying the same method as Theorem 9.6.1 to this cylinder, it is possible to construct k cylinders such that each of them can fold to a regular tetrahedron of the corresponding area (Fig. 9.25). Thus, we have the theorem. □

References

[Gol96] S.W. Golomb, *Polyominoes: Puzzles Problems, and Packings* (Princeton University, Patterns, 1996)

[Gar08] M. Gardner. *Hexaflexagons, Probability Paradoxes, and the Tower of Hanoi* (Cambridge, 2008)

[Gar14] M. Gardner, *Knots and Borromean Rings, Rep-Tiles, and Eight Queens* (Cambridge, 2014)

[ABD+17] Z. Abel, B. Ballinger, E.D. Demaine, M.L. Demaine, J. Erickson, A. Hesterberg, H. Ito, I. Kostitsyna, J. Lynch, R. Uehara, Unfolding and dissection of multiple cubes, tetrahedra, and doubly covered squares. J. Inf. Process. **25**, 610–615 (2017)

[Tor92] P. Torbijn, Cubic hexomino cubes, in *Cubism for Fun*, vol. 30 (1992), pp. 18–20

[Tor02] P. Torbijn, Covering a cube with congruent polyominoes, in *Cubism for Fun*, vol. 58 (2002), pp. 18–20

[Tor02b] P. Torbijn, Covering a cube with congruent polyominoes (2), in *Cubism for Fun*, vol. 59 (2002), p. 14

[Tor03] P. Torbijn, Covering a cube with congruent polyominoes (3), in *Cubism for Fun*, vol. 61 (2003), pp. 12–16

[XHU17] D. Xu, T. Horiyama, R. Uehara, Rep-cubes: unfolding and dissection of cubes, in *The 29th Canadian Conference on Computational Geometry (CCCG 2017)* (Ottawa, Canada, 2017), pp. 62–67

[Xu17] D. Xu, Research on Developments of Polycubes. Ph.D. thesis, School of Information Science, JAIST (2017)

[DO07] E.D. Demaine, J. O'Rourke, *Geometric Folding Algorithms: Linkages, Origami Polyhedra* (Cambridge, 2007)

[AHU16] Y. Araki, T. Horiyama, R. Uehara, Common unfolding of regular tetrahedron and johnson-zalgaller solid. J. Graph Algor. Appl. **20**(1), 101–114 (2016)

[Ser00] R. Séroul, 2.2. prime number and sum of two squares, in *Programming for Mathematicians* (Springer, Berlin, 2000), pp. 18–19

Chapter 10
Common Nets of a Regular Tetrahedron and Johnson-Zalgaller Solids

Abstract As we already saw in Sects. 4.2 and 4.3, common nets for two (or more) regular polyhedra are tough problems, and we cannot say anything about their existence. On the other hand, as introduced in Sect. 2.3, only for nets of a regular tetrahedron, its beautiful and useful characterization is known as a notion of $p2$ tiling. Then, what happens if one is limited to a net of a regular tetrahedron and the other is limited to an edge-unfolding of a more general polyhedron? In this chapter, as the latter one, we will investigate "an edge-unfolding of a regular-faced convex polyhedron".

10.1 Extension to Regular-Faced Convex Polyhedron

There are many kinds of regular-faced convex polyhedra. Therefore, in their edge-unfoldings, there may be a net of a regular tetrahedron. For this question, the answer is [Yes]. Specifically, the following theorem is known.

Theorem 10.1.1 *(1) Among the regular-faced convex polyhedra, a polyhedron called J17 solid has 13014 edge-unfoldings in total, of which 87 can be folded to regular tetrahedra. In particular, 78 of these 87 have only one way of folding to regular tetrahedra, but eight have two ways of folding and one has three ways of folding.*
(2) Among the regular-faced convex polyhedra, a polyhedron called J84 solid has 1109 edge-unfoldings in total, of which 37 can be folded to regular tetrahedra. In particular, 32 of these 37 have only one way of folding regular tetrahedra, but five have two ways of folding.
(3) Any edge-unfolding of a regular-faced convex polyhedron cannot be folded to a regular tetrahedron except a J17 solid, a J84 solid, and a regular tetrahedron,

That is, among the 5 regular polyhedra, 13 semi-regular polyhedra, Archimedean prisms and anti-prisms, and 92 types of Johnson-Zalgaller (JZ, for short) solids, only

R. Uehara, *Introduction to Computational Origami*,
https://doi.org/10.1007/978-981-15-4470-5_10

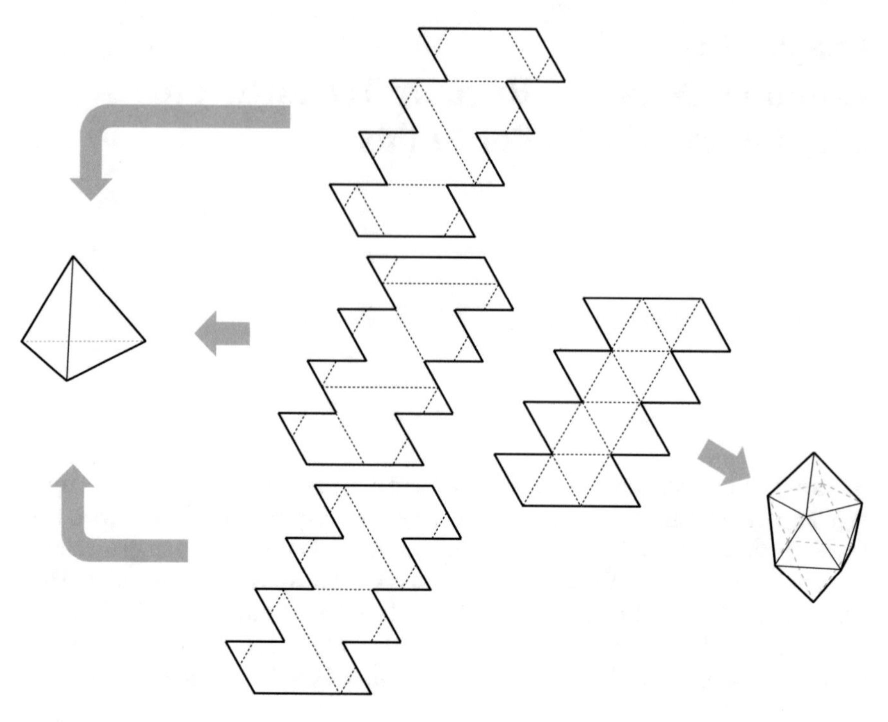

Fig. 10.1 The unique edge-unfolding of J17 that folds to a regular tetrahedron in three ways

J17 and J84 have edge-unfoldings that can fold to a regular tetrahedron.[1] Moreover, Theorem 10.1.1 claims that some of the edge-unfoldings of J17 and J84 solids have multiple ways for folding a regular tetrahedron. In Theorem 10.1.1 (1) and (2), examples of impressive "edge-unfoldings that can fold to a regular tetrahedron in multiple ways" are shown in Figs. 10.1 and 10.2. (The nets other than that shown here are on public at https://www.al.ics.saitama-u.ac.jp/horiyama/research/unfolding/common/.) The results shown in this section are basically extensions of the results in Sect. 4.2. We need more techniques to handle them efficiently since there are particularly many objective solids, and each solid has an astronomical number of nets. In the following, we will not give all details about every technique, but we will aim for an intuitive understanding of rough stories and techniques. For details, refer to [AHU15, AHU16].

[1] Of course, except for the net of a regular tetrahedron itself.

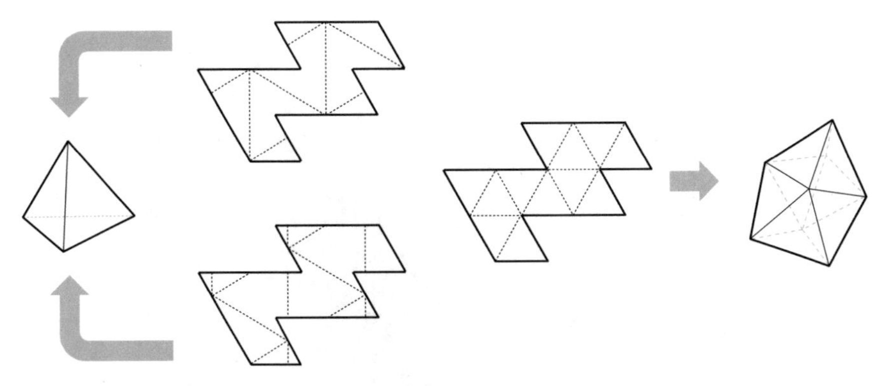

Fig. 10.2 One of five edge-unfoldings of J84 that fold to regular tetrahedra in two ways

10.1.1 Regular-Faced Convex Polyhedra Which Cannot Be p2 Tiling

First, due to some simple properties, there are many polyhedra that cannot be a p2 tiling in any edge-unfolding. For example, a regular dodecahedron is composed of 12 regular pentagonal faces, but as already discussed in Sect. 4.2, since the angles are multiple of unit angles 108°, any edge-unfolding of it cannot be a p2 tiling based on 180° rotation and copy. In a similar reason, any polyhedron in our category having a face of regular n-gon with $n > 6$ cannot have any edge-unfolding is a p2 tiling. A detailed proof can be found in [AKL+11], but the following theorem is already known.

Theorem 10.1.2 ([AKL+11]) *Of the polyhedra considered in this section, any polyhedron containing a regular n-gon that satisfies $n = 5$ or $n \geq 7$ cannot be a p2 tiling in any edge-unfolding.*

Therefore, considering the result of Sect. 4.2, at this point, the polyhedra where some edge-unfoldings may fold to a regular tetrahedron are the Archimedean hexagonal prism and anti-prism and JZ solids J1, J8, J10, J12, J13, J14, J15, J16, J17, J49, J50, J51, J84, J86, J87, J88, J89, and J90. In the following, the discussion will be complicated, so we will omit the discussion of Archimedean hexagonal prism and anti-prism, and concentrate on these JZ solids. Although we narrowed down a lot, there are still many things to investigate one by one. We consider a slightly more elaborate discussion.

10.1.2 Regular-Faced Convex Polyhedra Which Can Make p2 Tilings

We first note that, the almost polyhedra where some edge-unfoldings may fold to a regular tetrahedron satisfy the conditions in Theorem 2.3.2, and hence they have

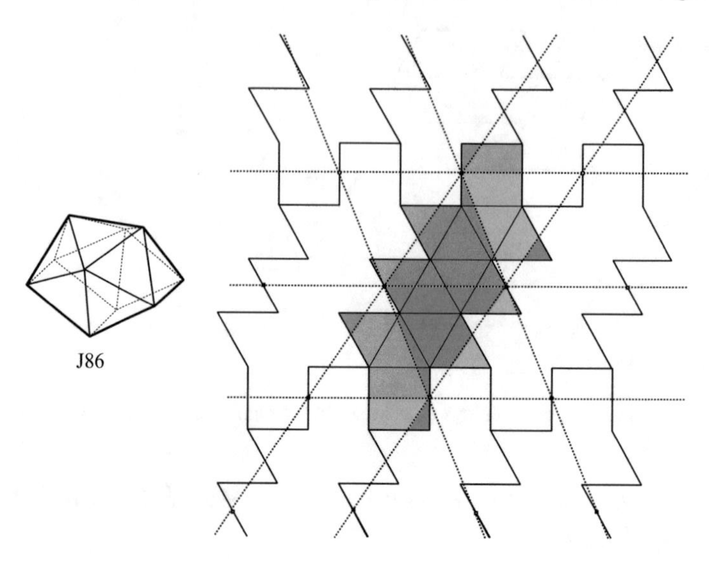

Fig. 10.3 A p2 tiling of J86. Folding along the dotted line, we obtain a tetramonohedron. Each face is colored to make it clear

some edge-unfoldings that fold to some tetramonohedra. Many concrete examples are shown in [AKL+11, AHU16]. For example, Fig. 10.3 shows an example of p2 tiling by an edge-unfolding of J86. That is, such an edge-unfolding can be folded to a tetramonohedron.

In Fig. 10.3, the crease lines are indicated by dotted lines and each triangular face is colored, but it is still quite difficult to imagine the tetramonohedron folded from this polygon with only this figure. On the other hand, it is easy to see that this polygon forms a tiling and that the point representing each vertex is the center of rotational symmetry. These facts tell us the power of characterization by a tiling. (Of course, if you actually cut out the polygon and assemble the tetramonohedron, you can immediately understand that the tetramonohedron can be folded.)

Exercise 10.1.1 Figure 10.4 is one of the edge-unfoldings of J89. Confirm that this is a p2 tiling and find one tetramonohedron that can be folded from this polygon.

In this way, we can fold various tetramonohedra from many edge-unfoldings. However, none of them can be a regular tetrahedron. To show this fact, we need more quantitative arguments. Here we focus on area and edge length.

All of JZ solids J1, J8, J10, J12, J13, J14, J15, J16, J16, J17, J49, J50, J51, J84, J86, J87, J88, J89, and J90 consist of only unit squares (of area 1) and regular triangles (of area $\sqrt{3}/4$). Let $S_1, S_8, S_{10}, \ldots, S_{89}, S_{90}$ be their areas, respectively. For each of them, assuming that there exists a regular tetrahedron with such a surface area, the length of an edge of the regular tetrahedron can be computed by the area. Let $L_1, L_8, L_{10}, \ldots, L_{89}, L_{90}$ be the lengths of edges, respectively. Table 10.1 shows all of them.

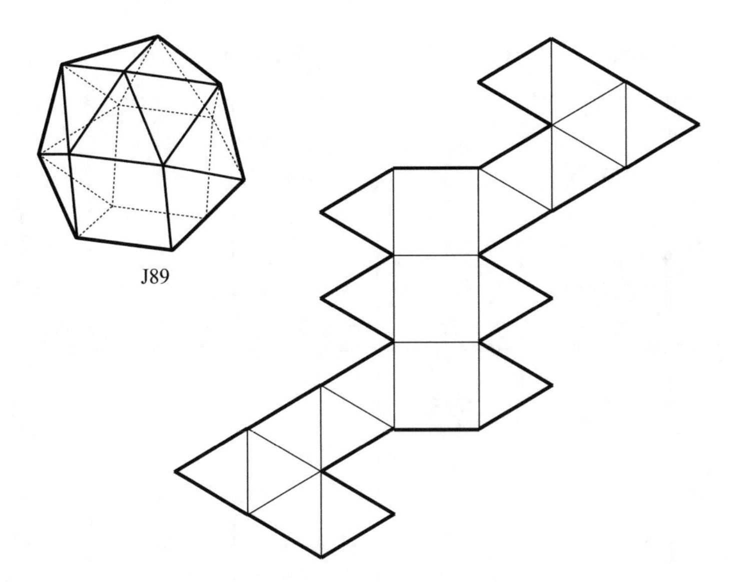

Fig. 10.4 An example of edge-unfolding of J89. It forms p2 tiling

Now, we consider an edge-unfolding of a certain JZ solid Ji. Let P_i be this polygon. The length of each edge of the polygon P_i is unit length. First, we generalize the arguments used in Sect. 4.2. This is an important property we will use, so we put it as a formal lemma.

Lemma 10.1.3 *Suppose that all edges of a polygon P have unit length, and a convex polyhedron Q can be folded from P. Furthermore, for every vertex of Q, the angle of the surrounding paper is assumed to be greater than* 180°. *Then any vertex v of Q has the following properties on P: (1) v is a point on the boundary of P. (2) v is at the end point or midpoint of a unit length edge of P.*

Proof Property (1) can be obtained immediately from Theorem 2.1.3. Thus, we focus on (2). First note that for property (2), a vertex v on Q may be generally scattered at several points on the boundary on P. These points are glued together so that the sum of their angles matches the angle around the vertex v on Q.

Here, the cutting lines on Q when unfolding to the net P from the convex polyhedron Q induce a spanning tree T of the vertices on Q by Theorem 2.1.1. We can also show that P is not convex by the same argument as Lemma 4.2.2. Therefore, there always exists a vertex v of Q that exceeds 180° on the boundary of P.

Now we consider an arbitrary subtree T' of T. That is, this T' gives a set of cut lines that appears in the middle process of unfolding of Q to P. This cut lines are not necessarily flat, but we can use the same argument as Lemma 4.2.2, and when we consider the leaves of T', there should be a point that exceeds 180° on the boundary. We call this property "non-convexity in the cut" for convenience.

Now we move on to proof of property (2). First, P is not convex, and hence there is a vertex v that exceeds 180°. When we fold the convex polyhedron Q from polygon

Table 10.1 The length L_i of an edge of the regular tetrahedron of the same area S_i of the JZ solid Ji

Solid	J1	J8	J10	J12	J13
Figure					
# of □s	1	5	1	0	0
# of △s	4	4	12	6	10
L_i	$\sqrt{\frac{\sqrt{3}}{3}+1}$ $= 1.255\cdots$	$\sqrt{\frac{5\sqrt{3}}{3}+1}$ $= 1.971\cdots$	$\sqrt{\frac{\sqrt{3}}{3}+3}$ $= 1.891\cdots$	$\sqrt{1.5}$ $= 1.224\cdots$	$\sqrt{2.5}$ $= 1.581\cdots$

Solid	J14	J15	J16	J17	J49
Figure					
# of □s	4	5	0	2	3
# of △s	8	10	16	6	6
L_i	$\sqrt{\sqrt{3}+\frac{3}{2}}$ $= 1.797\cdots$	$\sqrt{\frac{4\sqrt{3}}{3}+2}$ $= 2.075\cdots$	$\sqrt{\frac{5\sqrt{3}}{3}+\frac{5}{2}}$ $= 2.320\cdots$	2	$\sqrt{\frac{2\sqrt{3}}{3}+\frac{3}{2}}$ $= 1.629\cdots$

Solid	J50	J51	J84	J86	J87
Figure					
# of □s	1	0	0	2	1
# of △s	10	14	12	12	16
L_i	$\sqrt{\frac{\sqrt{3}}{3}+\frac{5}{2}}$ $= 1.754\cdots$	$\sqrt{3.5}$ $= 1.870\cdots$	$\sqrt{3}$ $= 1.732\cdots$	$\sqrt{\frac{2\sqrt{3}}{3}+3}$ $= 2.038\cdots$	$\sqrt{\frac{\sqrt{3}}{3}+4}$ $= 2.139\cdots$

Solid	J88	J89	J90		
Figure					
# of □s	2	3	4		
# of △s	16	18	20		
L_i	$\sqrt{\frac{2\sqrt{3}}{3}+4}$ $= 2.270\cdots$	$\sqrt{\sqrt{3}+\frac{9}{2}}$ $= 2.496\cdots$	$\sqrt{\frac{4\sqrt{3}}{3}+5}$ $= 2.703\cdots$		

P, we cannot glue other edge to this vertex v. Since it exceeds 360° as a whole, it contradicts that Q is convex. Then there are two cases.

Case 1: The vertex v is glued alone. This means that the two edges incident to v are glued to each other. In this case, the edges of unit length on both sides must be completely glued. Otherwise, if another piece of paper is inserted in between them, the angle at that part exceeds 360°, which contradicts the convexity of Q. As a result of completely gluing the edges of unit length, the vertices on both sides of v are

glued to become a new vertex w. If the angle at w is more than 180°, we replace v with w and proceed inductively with the number of vertices. Even if the angle at w is less than 180°, if the cut still remains and Q has not been completed, we can proceed inductively from the above "non-convexity in the cut" property.
Case 2: Another vertex v' is glued to the vertex v. We note that two or more vertices can be glued together. In any case, as in Case 1, edges of unit length are usually glued. Then the cut on Q is divided into multiple sections, but we can proceed inductively for each cut.

Therefore, in any case, it can be basically considered that gluing is performed for each pair of two edges of unit length. The only exception is when gluing adjacent vertices. In this case, a vertex of Q is also generated at the midpoint of an edge of unit length.

Summarizing the above arguments, it can be seen that the points on the boundary of P where a vertex of Q can appear are limited to the end point or midpoint of an edge of P of unit length. □

Now we apply Lemma 10.1.3 for Ji and P_i. The assumption is that the regular tetrahedron is the convex polyhedron folded from P_i. In this case, we have to pay attention to convexity and non-convexity. The paper around the apex is 180°, and otherwise the paper is 360° in the case of a regular tetrahedron, and P_i is a net of a convex polyhedron Ji that originally satisfies the condition of Lemma 10.1.3. Therefore, the claim in Lemma 10.1.3 still holds. Specifically, by Lemma 10.1.3, if a regular tetrahedron is folded from P_i, each of all four vertices of the tetrahedron can exist only at the end point or middle point of the edge of unit length of P_i.

That is, to summarize the current situation, the following two hold.

- P_i is a non-convex polygon formed by joining some regular triangles and some squares, where each edge is unit length.
- The end points of one edge L_i of a regular tetrahedron are the end points or midpoints of the unit length edges of P_i.

Now we revisit the powerful technique "commutability of vectors" that appeared in Sect. 4.2. From the above discussion, L_i is the sum of several 1/2-long vectors, and each vector changes direction only at multiples of 30°. Currently, only the positions of the start and end points of L_i are in question, so the commutative property of vectors implies that even if these vectors are rearranged in order of orientation, it is not the problem since the point to finally reach does not change. Let $\vec{L_i}$ be this standardized vector obtained from L_i. Then $\vec{L_i}$ can be represented as

$$\vec{L_i} = \frac{k_1}{2}(1, 0) + \frac{k_2}{2}(\cos 30°, \sin 30°) + \frac{k_3}{2}(\cos 60°, \sin 60°) + \frac{k_4}{2}(0, 1)$$

for some positive integers $k_1, k_2, k_3,$, and k_4. Thus, the JZ solids with a double root appearing in L_i, specifically J1, J8, J10, J14, J15, J16, J49, J50, J86, J87, J88, J89, and J90 are immediately rejected. Therefore, only J12, J13, J17, J51, and J84 remain.

Finally, we use the power of a computer to solve it. Specifically, we check all positive integers k_1, k_2, k_3, and k_4 such that $\left|\vec{L_i}\right|$ is less than 2. Then we can immediately confirm that lengths other than J17 and J84 cannot be realized as vectors. From the above, it was proved that there is no solution other than J17 and J84.

The power of vector

The contents in this section are based on the contents of the journal paper [AHU16]. When these results were first presented at an international conference [AHU15], we did not come up with the idea of using the vector. Thus, we first considered all zigzag lines without self-crossing induced by unit squares and unit regular triangles, gave the upper bound of the length of such zigzag lines, and checked all possible combinations by a computer program. The computation time in this way required about 10 h. However, when we were rewriting to publish the article in the journal, we realized that using the vector can dramatically speed up. Although the program rewrite was only a few lines, the computation time was actually reduced to less than *one second.* It succeeded in speeding up roughly a few tens of thousands times. We can say that the improvement of the algorithm sometimes leads to dramatic efficiency improvement.

10.1.3 Regular-Faced Convex Polyhedra Having Common Developments with Regular Tetrahedron

Finally, we show that there are edge-unfoldings of J17 and J84 that can fold to regular tetrahedra. As shown in Theorem 10.1.1, J17 has 13014 edge-unfoldings, of which 87 can fold to regular tetrahedra. On the other hand, J84 has 1109 edge-unfoldings, of which 37 can fold to regular tetrahedra. Moreover, in some of the nets where they can fold to regular tetrahedra, there are multiple ways of folding.

In order to show these facts, we have to solve two problems. Specifically, (1) enumerating all edge-unfoldings of a given polyhedron and (2) examining whether or not we can fold to a regular tetrahedron from a given edge-unfolding, and how many ways if we can.

It will not be obvious what kind of algorithms we have to use to solve these problems. In the following, we will not limit the targets to J17 and J84 considered here, and give an outline of how to solve these problems in more general form.

10.2 Enumeration of All Edge-Unfoldings of a Given Convex Polyhedron

We first consider the problem of enumeration of edge-unfoldings of a given convex polyhedron Q (in the case of this problem, J17 and J84). A useful property here is Theorem 2.1.1. That is, if the vertices and edges of a convex polyhedron are regarded as a graph, an edge-unfolding of Q is given by a spanning tree. Therefore, we regard the vertices and edges of Q as a graph, and enumerate all spanning trees of the graph. The problem of enumeration of spanning trees of a given graph is a theme that is well studied in the context of graph theory and graph algorithms.

However, we have to note that this "edge-unfolding" is just giving the way of unfolding. Up to this point, we have just mentioned the enumeration of the way to open the edges of the polyhedron. We have not yet considered whether the faces of the polyhedron do not overlap. As mentioned in Sect. 2.2, there is currently no universal way to find out if the faces overlap when the polyhedron is unfolded. Also, if the polyhedron has symmetry, it is natural to consider all nets that will have the same shape when rotated or flipped as "same net". There are some general methods to check such duplications, as shown in the following column, but it is necessary to devise each problem.

Check for symmetry

In geometric enumeration problems, polygons that are identical with respect to rotation or flipping are often regarded as essentially duplicates of the same thing, and we often want to output only one. A common way is to "predetermine some order and output only the smallest one". For example, consider recording a polyomino in a two-dimensional array $p[i, j]$. Given the coordinates (i, j) inside a rectangle of size $n \times m$, we store 1 when there is a polyomino and 0 when there is not. Then this $p[i, j]$ can be thought as a binary number specific to that polyomino representation. Specifically, for example, we can compute it as

$$\sum_{0 \le i \le n-1, 0 \le j \le m-1} p[i + j \times m] \times 2^{i+j \times m}.$$

For a given polyomino, there can be eight different representations in total, considering the initial arrangement and all the other ways of rotations and flipping. Then all of eight numbers represent essentially the same polyomino. Therefore, the minimum of these eight values is regarded as the regular representation of this polyomino. When the algorithm enumerates all polyominoes, if one representation is found, it generates all eight symmetric forms, and duplication can be prevented by outputting only when the representation found the minimum value.

Thus, in order to enumerate all the edge-unfoldings of a given polyhedron, after creating one spanning tree, it is necessary to confirm that it has no overlap when it is cut and unfolded, and check the symmetry to avoid duplicates. The computation in this part is a bit troublesome. Moreover, since the ways of edge-unfoldings increase exponentially if it becomes a slightly complicated polyhedron, the enumeration problem of all the edge-unfoldings of a given convex polyhedron is a considerable challenge in practice.

A type of data structure called *Binary Decision Diagram (BDD)* as a technique to handle efficiently such explosively increasing data is actively studied in recent years (as introduced in Sect. 3.4.1). Conceptually, it is a simple binary decision tree, and it is a data structure that can be compressed by sharing a common structure when storing it as data. The binary decision tree is a natural structure that can be found in a kind of divination, for example, and can be summarized as follows.

Binary decision tree: Imagine assigning [Yes/No] to some choices in order. This is represented as a rooted binary tree structure that branches according to which one is assigned. This is a *binary decision tree*.

Binary decision diagram: A binary decision tree sometimes has exactly same structures. In this case, we can unify all of them into one common structure. In other words, in a binary decision tree, every vertex except the root has only one incoming edge, but in a binary decision diagram, sharing partial structures, some vertices can have multiple incoming edges. Because of this, the whole structure is no longer a tree. If the same structure appears many times as substructures, it will be a much more efficient data structure than a binary decision tree.

Though the idea is simple, BDD is very powerful. However, it does not always work well. The part of "assign to some choices in order" is the point. Finding a good order may improve memory efficiency exponentially, but there is no general method (under the assumption of $P \neq NP$) to find a good order efficiently. For more details on BDD, refer to [SB15].[2]

As introduced in Sect. 3.4.1, in the research of enumeration of nets, research using BDD has advantage. Within the algorithms for edge-unfoldings, the set of nets is managed by the following data structure:

1. A number is assigned to each edge of a polyhedron starting from 1.
2. Set the parameter "cut/not cut" to each edge.
3. Represent the set of spanning trees that consist of cutting edges (and non-cutting edges) as a BDD.

This method of representation allows us to manage the astronomical number of edge-unfoldings in a compact way.

Once the above BDD representation is completed, it is sufficient to output the correct edge-unfoldings in order and check the existence of overlap and the symmetry, but this process requires the different techniques for each polyhedron. Thus, we conclude this topic here in this book.

[2]See also the cite of Graphillion at http://graphillion.org which is a fast, lightweight library of BDD for a huge number of graphs.

10.3 How to Fold Convex Polyhedra from a Given Polygon

As already mentioned, except in regular tetrahedra, the relationship between polygons and convex polyhedra that can be folded from them is generally unknown. When there is no mathematical characterization, there seems to be no better way than to try all possible ways of folding by a computer. However, what does this *all* mean? For example, as seen in the rolling belt considered in Sect. 2.3.2, a computer cannot handle them if there are an infinite number of points that can be vertices. The general approach is not known and some partial results are known depending on the situation. Thus, we introduce some of the results known at the moment.

Let P be a given polygon and Q be a polyhedron to be folded. The shape of a general polyhedron is not determined uniquely. For example, you can make a different solid by pushing the pointed end of a cone into it. That is, you can obtain variations of a polyhedron with many bumps by inverting the bumps one by one. On the other hand, if it is limited to convex polyhedra, as introduced briefly by Theorem 7.2.3, once the geometrical relationship is determined, its shape is uniquely determined. In the current context, if there is a convex polyhedron Q that folded from P in a way, its shape is uniquely determined. Under such circumstances, it is convenient to assume that Q is a convex polyhedron. Thus, we assume that Q is a convex polyhedron hereafter.

10.3.1 Folding Regular Tetrahedron

We assume that the polyhedron Q to be folded is a regular tetrahedron (this is the problem we are considering in the first place). In this case, Theorem 2.3.2 can be used, hence it can be considered whether a given polygon P is a p2 tiling. For the problem we are considering now, there are some more conditions that can be used.

First, the polygon P was an edge-unfolding of J17 or J84. Thus, we can use Lemma 10.1.3. The polygon P is an equilateral polygon, not convex, and all vertices of Q are endpoints or midpoints of edges of these edges of P of unit length. Moreover, due to area constraints, the length of an edge of Q is 2 for J17 and $\sqrt{3}$ for J84. In both cases, it is not so difficult to find that one edge of Q can connect either two endpoints or two middle points of P. Therefore, it is sufficient to check whether each endpoint or midpoint on P can be a vertex of Q that satisfies the conditions listed here.

In the literature [AHU16], we represent the polygon P by a sequence of turn angles that change direction every 1/2 unit length, and check the conditions one by one. In the problem in [AHU16], since P is an equilateral n-gon, there are only $2n$ points that can be the rotational center, and n is not so large in this case, and the computational cost of this iteration is not a serious problem.

Moreover, in the tiling in this case, the following two conditions in a triangular grid of specified size have to be satisfied: (1) align a grid point with the center of rotational symmetry considered now, and (2) other grid points are on the boundary of P. Therefore, we first place P so that these two conditions are satisfied, and then check whether the grid points on the boundary of P are at the centers of rotational symmetry. The details of the specific algorithm are omitted because they become somewhat complicated, but we can check if it is p2 tiling by a relatively straightforward algorithm. If the data size is this large, the real running time is not a problem.

10.3.2 Folding Box

Next, we consider that the polygon Q we want to fold is a box of size $a \times b \times c$. When we know that P is a polyomino, a, b, c are integers, and it is supposed to fold along the grid lines of P, as we showed in the first half of Sect. 3, we can use the simple algorithm that virtually pastes P on Q.

Even if a, b, and c are integers, P may not be folded along grid lines due to area constraints. In that case, it is better to use the dual graph model as shown in Sect. 3.4.1. That is, it is a method of representing P by a lattice graph with a geometrical structure, and putting it in order on a lattice graph induced from Q to check whether overlaps or gaps occur. The geometric structure here means that each vertex distinguishes four directions of upper, lower, left, and right. Each distance in a pair of adjacent vertices is a unit length, and if the orientation is consistent, we can decide it correctly. Intuitively, we imaginary overlap a square grid of wire mesh on a box made of another wire mesh. This is also a relatively simple method. On the contrary, it seems difficult to construct an algorithm other than such a naive method.

In a recent paper by Mizunashi et al. [MHU19], we propose an algorithm for more general a, b, and c, which are not necessarily integers. This is a pseudopolynomial time algorithm that determines, for a polygon P of n vertices, if a box can be folded from P, and it finds all the boxes folded from P if they exist. As shown in Sect. 3, we may have several boxes folded from P, and for example, P can be a shape of a long ribbon that wraps Q in a non-trivial way. It is not as simple as it looks to fold a box by a polygon in this general form. Since this algorithm requires more knowledge of computational geometry and some techniques of algorithms, we omit it in this book.

10.3.3 Folding General Convex Polyhedron

There is not much known when P is a general polygon and Q is a general convex polyhedron. However, as introduced in Fig. 1.2 in Sect. 1, we know that the polygon called Latin cross, which is one of the 11 edge-unfoldings of a cube, can fold to 23 different polyhedra by 85 different ways of folding [DO07]. How did they find so many ways of folding and how did they show that they were all? We give an answer to this natural question at last.

First, this Latin cross P is an equilateral polygon and not convex. Thus, we can use Lemma 10.1.3, and the vertices of a convex polyhedron Q folded from P are all endpoints or midpoints of the unit length edges of P. Therefore, there are 28 candidate points that can be vertices of Q on these 14 edges of P.

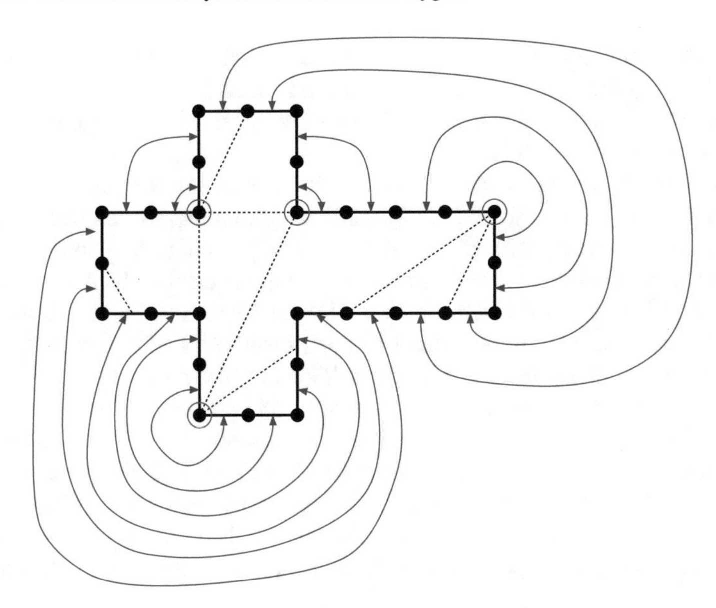

Fig. 10.5 Gluing of edges to fold a tetrahedron from the Latin cross

We consider what happens when a convex polyhedron Q is folded from this P. Since Q is a closed polyhedron, an edge of P should be glued to the other edge. We call the correspondence between pairs of edges a *matching*. We here note that this matching should be considered with 1/2 unit length. That is, P can be considered as a 28-gon virtually, and in this case, the length of each edge is exactly 1/2. Each of these half-length edges is glued to the other matching edge on Q. An example is shown in Fig. 10.5. In the figure, • is a virtual vertex, and the length of each edge is 1/2. If the matching between these edges is twisted, Q will not be a polyhedron. Intuitively, imagine starting from some vertex and gluing these edges in order. In this matching, there exists a pair of adjacent edges that they are glued each other. They are indicated by a red circle in the figure. These pairs make vertices. We consider to glue the corresponding edges in order from these vertices. Then, a polyhedron cannot be folded correctly if the order of the edges is twisted in this sequence of gluing. In other words, gluing of these edges should be in a nested structure without intersection. The correspondence of this gluing is shown by the blue arrows in the figure. That is, gluing of the edges based on such "non-crossing matching" is necessary to fold a polyhedron Q from a polygon P.

Once the matching is fixed, we can find the vertices of Q by adding the angles of the paper gathered around at each vertex on P. More precisely, it is flat if the total angle is 360°, and it becomes a vertex on Q if the total angle is less than 360°. Of the vertices in the figure, those less than 360° are red circled. Specifically, the angle at two vertices is 270°, and the angle at the other two vertices is 90°. According to Alexandrov's theorem, the shape of the convex polyhedron Q is uniquely determined by this.

Note that although one vertex on P forms one vertex of Q in this example, this is not generally the case. For example, when we fold a cube in the usual way from P in the figure, some of the vertices of Q will be formed in an angle of 270° with three vertices of 90° together.

Then how can we find such a matching? It takes an exponential time to list all possible matchings naively, but with dynamic programming, we can find all gluings that can make convex polyhedra in $O(n^3)$ time. Dynamic programming is one of the basic techniques of algorithm theory and is a powerful technique. Intuitively, computation is performed locally for each limited range, and the results are stored in a table (or an array). In the current context, for a series of consecutive vertices on the boundary of P, all local feasible foldings in that range are examined, and the ways of convexly folding in that range are kept in a table. Then, by expanding the range one by one, the table is gradually expanded. When we get to the end, we should have essentially all the foldings of the convex polyhedra preserved in the table. This technique is the same as the method of computing the bumpy pyramid with the maximum volume in Theorem 7.8.2. For the details of the specific algorithm to compute the gluing correspondence by dynamic programming, refer to the literature [DO07].

Now, the correspondence between the edges is decided, and the positions of the vertices are fixed. Then, can we say that the polyhedron Q is completely determined? The answer is [No]. Specifically, the crease lines for folding P to Q (indicated by dotted lines in the figure) are not yet known. Unfortunately, there is nothing more we can do. We do not know the general way to find these crease lines. According to the authors who proved that the result of "23 different convex polyhedra can be folded from the Latin cross in 85 different ways", this last step is confirmed by their hands.

Exercise 10.3.1 In Sect. 10.3, we introduced that the Latin cross P can fold to 23 different convex polyhedra Q. In the argument, we used Lemma 10.1.3 since Q is *convex*. Then what happens if we allow to fold to non-convex polyhedra as Q? Actually, an infinite number of non-convex doubly covered polygons Q can be folded from P. Consider such Q.

We here give a brief summary to make sure:

- When we fold some polyhedron Q from a given polygon P, the pairs of edges for gluing have nested structure. After gluing, if each angle around each vertex is less than or equal to 360°, Q forms a convex polyhedron.
- We can find all gluings that fold to convex polyhedra in $O(n^3)$ time using dynamic programming.
- We can find all the vertices of Q and the way of gluing for a feasible gluing.
- However, there is no general way to compute where the edges on Q, that is, where the internal crease lines of P.

The last step is a very annoying problem. Note that in the petal folding problem considered in Sect. 7, Q is limited to a bumpy pyramid, and we reduce the original problem to the other problem for finding a "good" triangulation of the bottom n-gon. Here is a field where future research is desired.

Open Problem 10.3.1 *Given a polygon P with the correspondence between edges of P for gluing, find a (more) general method for finding the crease lines of P to fold a convex polyhedron Q.*

References

[AHU15] Y. Araki, T. Horiyama, R. Uehara, common unfolding of regular tetrahedron and Johnson-Zalgaller solid, in *The 9th Workshop on Algorithms and Computation (WALCOM 2015)*. Lecture Notes in Computer Science, vol. 8973 (2015), pp. 294–305

[AHU16] Y. Araki, T. Horiyama, R. Uehara, Common unfolding of regular tetrahedron and Johnson-Zalgaller solid. J. Graph Algorithms Appl. **20**(1), 101–114 (2016)

[AKL+11] J. Akiyama, T. Kuwata, S. Langerman, K. Okawa, I. Sato, G.C. Shephard, Determination of all tessellation polyhedra with regular polygonal faces, in *The 9th International Conference of Computational Geometry, Graphs, and Applications (CGGA 2010)*. Lecture Notes in Computer Science, vol. 7033 (2011), pp. 1–11

[SB15] T. Sasao, J.T. Butler (eds.), *Applications of Zero-Suppressed Decision Diagrams* (Morgan & Claypool Publishers, 2015)

[MHU19] K. Mizunashi, T. Horiyama, R. Uehara, Efficient algorithm for box folding, in *The 13th International Conference and Workshop on Algorithms and Computation (WALCOM 2019)*, Lecture Notes in Computer Science, vol. 11355 (2019), pp. 277–288

[DO07] E.D. Demaine, J. O'Rourke, *Geometric Folding Algorithms: Linkages, Origami, Polyhedra* (Cambridge, 2007)

Chapter 11
Undecidability of Folding

Abstract The concluding chapter of this book is the topic of origami modeling. So far, we mainly consider discrete origami problems, which suit computers. However, when we consider continuous problem on origami, we have to face a gap between discrete and continuous models. Using this gap, we can consider undecidability on origami.

11.1 Theoretical Model of Origami

The field of "computational origami" is related to "computation", and hence there are many suitable researches for using a computer. In short, many models have discrete structures. In most cases, integers are often enough to handle data, and the problem of computational error does not have to be concerned.

However, a real sheet of paper is not discrete, but continuous and seems to be folded anywhere. When dealing with such continuous quantities, it is necessary to decide the model carefully. If you think that the sheet of paper is quite large, the thickness of the paper is zero, and it can be folded with any accuracy, there are actually many strange things.

For example, if you repeatedly fold the paper in half, it will increase to 2, 4, 8, 16 sheets in total. In an ideal model, this can be repeated as many times as you like, but if you actually fold a newspaper, you cannot repeat even ten times. If you fold the newspaper 10 times in half, the number of sheets becomes $2^{10} = 1024$, and even if the thickness of the newspaper is 0.1 mm, the thickness becomes roughly 10 cm. This is probably impossible.[1]

Aside from the unnaturalness associated with such physical idealization, delicate issues are hidden in the place "with any accuracy". For example, given an origami of size 1×1, it is easy to make a line segment of length $\sqrt{2}$: Just fold along the diagonal line. So what if you are asked to fold at $1/\pi$ in length? The ratio $\pi = 3.14592\cdots$

[1]According to the Guinness World Records (https://www.guinnessworldrecords.com/world-records/494571-most-times-to-fold-a-piece-of-paper), a high school student, Britney Gallivan, folded a single piece of paper in half 12 times on January 27, 2002. However, she used the tissue paper of 4,000 ft long. So I bet nobody can fold a newspaper in half 10 times.

R. Uehara, *Introduction to Computational Origami*,
https://doi.org/10.1007/978-981-15-4470-5_11

is an infinite number of decimal places. How can you fold such lengths? In general, how can we deal with the problem of folding a given *arbitrary* length?

The purpose of this chapter is to give some answer to this problem. It is a rather subtle topic of theoretical computer science, and it may be difficult for readers who are not familiar with such a theme. Therefore, we first introduce the topics of "diagonalization" and "halting problem" of a computer, which seemingly unrelated, and then consider the undecidability of origami folding. Readers who are accustomed to these topics may skip these sections, or readers who shudder just hearing diagonalization may not enjoy this chapter.

11.2 Diagonalization

We first explain the diagonalization briefly. This is a technique devised by mathematician Cantor to deal with two infinities, "countable" infinite and "uncountable" infinite. This is a kind of contradiction method, but it is also a technique that plays an important role in the field of theoretical computer science as shown in the next section.

We now consider two infinite sets, say A and B. Then the numbers of elements included in A and B are called *cardinalities* of them. (This is a mathematical term introduced because it would be strange to call an infinite number of things as "number", so it is not needed to worry about this word.)

Then, how can we compare the "sizes" or "cardinalities" of A and B? The great idea by Cantor is a sort of decision: we consider two cardinalities are regarded as the same if there is a one-to-one correspondence. Thus, for example, all the following sets have the same cardinality:

Natural numbers: Here, let us say that the set N of natural numbers is defined by $N = \{0, 1, 2, \ldots\}$. Although there is a style that does not include 0, I like to include 0.

Non-negative integers: Let $N^+ = \{1, 2, \ldots\}$ be a set of positive integers. If we define $f_1 : N \rightarrow N^+$ as $f_1(x) = x + 1$, we will have a natural one-to-one correspondence. So N and N^+ have the same cardinality. We do not consider that N^+ has less item than N.

Even numbers: Let $E = \{0, 2, 4, 6, 8, \ldots\}$ be a set of non-negative even numbers. If we define $f_2 : N \rightarrow E$ as $f_2(x) = 2x$, we have a natural one-to-one correspondence. In other words, N and E have the same cardinality, and it is not the case that E is only half of N.

Primes: Let $P = \{2, 3, 5, 7, 11, 13, \ldots\}$ be a set of prime numbers. We define $f_3 : N \rightarrow P$ as "$f(x) = p$, where p is the xth prime number". Since this is also a natural one-to-one correspondence function, N and P have the same cardinality.

Let us add a little more about the final P. First, it is known that there are infinite prime numbers.[2] Also, since the prime numbers can be enumerated in ascending order, the term "xth prime number" can be defined exactly in the mathematical sense. Thus, f_3 is a so-called "well-defined" function. Readers who are not familiar with these arguments may feel uncomfortable. In particular, we have no idea (so far) how we can compute the function f_3 specifically. The algorithm of computing the function f_3 is not a problem here. If it can be shown that a function with the property does exist, that is enough.

Looking at it this way, any infinite set that comes to mind naturally can make a one-to-one correspondence with the natural number. In other words, this is equivalent to that the set can be ordered as the first element, the second element, and so on. Namely, every infinite set that has the same density as a natural number can be ordered. This is a set with a very natural property in a sense, so it is called *countable set*.[3] This means that we can count the set from the smallest elements in order.

Then, is there any infinite set that is "uncountable" or "unordered"? Actually, it exists. The set of real numbers cannot be counted. Here, for the sake of simplicity, let us show that even a very limited interval among real numbers, specifically, real numbers included in (0, 1) are not countable.

Theorem 11.2.1 *Let R be the set of real numbers contained in the interval* (0, 1). *Then R is an uncountable set.*

Proof In order to derive a contradiction, we assume that R is a countable set. When R is a countable set, by the definition of countable, there exists a function $f : N^+ \to R$ that gives a one-to-one correspondence with the set N^+ of non-negative integers. That is, it can be listed as the first element r_1, the second element $r_2, \ldots,$ ith element r_i, ... Now r_i is a real number contained in the interval (0, 1), so it can be expressed as

$$r_i = 0.r_{i,1}r_{i,2}r_{i,3}\cdots r_{i,j}\cdots ,$$

where each $r_{i,j}$ is the jth decimal number in r_i. That is, $r_{i,j} \in \{0, 1, 2, \ldots, 9\}$.

Then we define the new number x as follows:

$$x = 0.x_1x_2x_3\cdots x_j\cdots ,$$

where x_j is defined as $x_j = 5$ if $r_{j,j} = 3$ and $x_j = 3$ if $r_{j,j} \neq 3$ (Fig. 11.1).

Here, by definition, the number x is a real number because each jth digit x_j is well defined. Thus $x \in R$. We assume that R was countable. Therefore, there is a natural number k such that x is the kth element of R. Now, consider the kth digit of x. It is either $x_k = 3$ or $x_k = 5$ by the definition of x. However, if $x_k = 3$, then

[2]This is an example that often appears as an introduction to the contradiction method, but let us introduce it briefly. Assuming that there is only a finite number of prime numbers. Then "the number multiplied by all prime numbers" +1 is a new prime number. This is a contradiction.

[3]Finite sets are also called *countable* sets. A finite set cannot make a one-to-one correspondence with N, but it can be arranged in order.

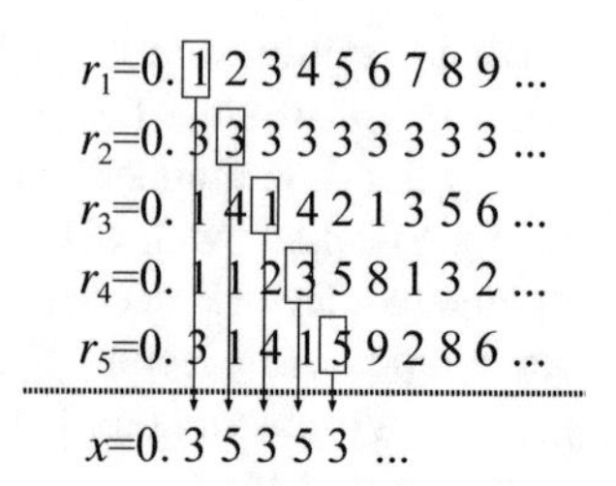

Figure 11.1 Definition of a new number x

$r_{k,k} = 3$, and by definition x_k should be 5. This is a contradiction. On the other hand, if $x_k = 5$, then $r_{k,k} = 5$, so we have $x_k = 3$, which is also a contradiction. Thus, the assumption that R is a countable set is not true. □

This proof technique is called diagonalization because we construct the number x from the digits that appear along the diagonal when we write $r_{i,j}$ on a plane. Readers who are new to diagonalization should consider carefully.

11.3 Undecidability of Halting Problem

Here, although it is somewhat out of the scope of this book, let us take up the halting problem as an interesting application of diagonalization. The essence of the halting problem is that there are functions that cannot be computed in principle even on a computer. Yes, there are functions that cannot be computed *in principle* on a computer. To put it another way, there are certain problems such that we cannot make any computer programs for solving them in principle. In this section, we prove that such problems exist. The person who never heard the proof of the undecidability of the halting problem would be surprised: "why can such a thing be proved mathematically?" If you look at the proof carefully, essentially, it is a diagonalization, and the logic itself is not so difficult (if you accept diagonalization). However, the meaning of the result is deep.

In theoretical computer science, a Turing machine is often used as a computation model. Although there are several variants of computation models which were devised independently one after another around 1940s, they all have essentially the same computation ability. This is the same with any programming language. Thus, we decide one programming language the readers like. Let us use the programming language C, for example, according to my preference. (As you can see below, this language can be any programming language. You don't need to know C.) Programs written in C are written in a fixed alphabet, i.e., alphanumeric characters and symbols, and have a finite length. This alphabet is expressed as binary data in a computer. In short, any computer program is represented by a string of 0 and 1 s of finite length. For the sake of distinction, we write the program itself in uppercase letters and the data given to it in lowercase letters. For simplicity, let us denote the output of a program P with a given data x by $P(x)$. For example, $P(x) = 1$ means that 1 is output on the screen when data x was given to program P. One important point here is that $P(x)$ can get into an infinite loop and does not produce any output. This special situation is denoted by $P(x) = \perp$.[4]

Here such a loose framework is enough, and any computer language has a common computation power with the same limitations. Now we are ready. The problem to be considered here is the following *halting problem*:

Input: Program P and input x.
Output: Does $P(x)$ terminate in finite steps when P is executed with the given input x? In other words, $P(x) \neq \perp$?

The seemingly simple decision problem is an undecidable problem. That is, there is no program Q that always returns the correct answer for this problem in finite time.

Can't we build a universal debugger?

Some readers who have made programs may have used software called *debugger*. A debugger is software for analyzing the behavior inside a program that has an unintended execution. In that sense, you will want the debugger to determine if your program terminates or not. The above claim shows that it is impossible in principle. Precisely, software that determines termination for *any* program cannot be made in principle (as we will show from now on). However, in the real world, it is not necessary to be so pessimistic. For example, you can make software that determines termination for *most* programs. The area covered by this "most" is up to the skill of programmer who make the debugger.

Hereafter we will prove this using diagonalization. Here is a preparation for that:

Lemma 11.3.1 *There are countable infinite number of programs written in C.*

Proof It is sufficient to show one way to enumerate the programs. We first introduce a lexicographic order between two strings P and P' as follows:

- If the lengths of P and P' are different, the shorter is smaller.
- If the lengths of P and P' are the same, we use an ordinary lexicographic order.

Specifically, for the alphabets 0 and 1, they are arranged in the following order.

$$\epsilon, 0, 1, 00, 01, 10, 11, 000, 001, 010, 011, 100, 101, 110, 111, 0000, 0001, \ldots,$$

where ϵ is an empty string of length 0. As mentioned above, the program P written in C is ultimately recorded inside the computer as a string p consisting of 0 and 1 s.

[4] This symbol "$\perp$" is called *bottom*, and it is customary to use it in the society of theoretical computer science. I do not know the historical origin.

Therefore, of course, it should eventually appear in the strings of 0 and 1 s shown above. Conversely, the sequence of 0 and 1 s above is a mixture of strings that are valid expressions and not valid expressions of C programs in order. If you extract only those that are valid expressions of C programs and arrange them, you can assign indices to the first program, the second program, and so on in order. This fact shows that the set of valid C programs is countable. Also, if you add artificially unnecessary descriptions and extend them, you can create as long a program as you like, so there are countable infinite number of C programs.

Now we turn to the main theorem in this section.

Theorem 11.3.1 *There is no program written in C that solves the halting problem.*

Proof We prove it by contradiction. Suppose that there is a program Q that solves the halting problem. For a given C program P and an input x to P, this Q outputs 0 if $P(x) = \perp$ and 1 if $P(x) \neq \perp$. That is, when $P(x)$ terminates anyway, it does not care what output it can get, but outputs 1 anyway. Formally, if the binary representation of P is p, it can be written as follows. (Here, $< p, x >$ means that program P is represented in a binary string p, and this is paired with the binary input x and expressed as a binary string as a whole, but you don't need to care too much.)

$$Q(< p, x >) = 0 \text{ if } P(x) = \perp,$$
$$Q(< p, x >) = 1 \text{ if } P(x) \neq \perp.$$

Now, looking at the program code Q, there should be operations that output 0 and operations that output 1. We rewrite this part a little and make a new program Q'. The input is the same as Q, the C program P, and its input x. However, the output is slightly different. Specifically, it is as follows:

$$Q'(< p, x >) = 1 \text{ if } P(x) = \perp,$$
$$Q'(< p, x >) = \perp \text{ if } P(x) \neq \perp.$$

That is, rewrite all operations that output 0 to output 1, and rewrite all operations that output 1 in an infinite loop. Infinite loops are easy to write in any programming language. The C language is not an exception. Thus, Q' also exists if Q exists.

By Lemma 11.3.1, there are countable infinite number of programs written in C. That is, they can be listed as $P_1, P_2, P_3, \ldots$. In other words, these programs are represented by the binary strings $p_1, p_2, p_3, \ldots$ such that $p_1 < p_2 < p_3 < \cdots$. Now consider $P_i(p_j)$ for any positive integers i and j. That is, $P_i(p_j)$ is the output of the ith program P_i when the binary representation of the jth program P_j is given as input data to P_i. A table of the output is shown in Fig. 11.2. This table shows all the programs written in C (in finite length). Therefore, Q' should appear somewhere in this table. Let k be the index of Q', that is, we suppose that $Q' = P_k$. Now we consider $P_k(p_k)$. Then, due to the construction of Q', either $P_k(p_k) = \perp$ or $P_k(p_k) = 1$. If $P_k(p_k) = \perp$, Q' should output 1 and stop at this time according to the definition of $Q'(< p, x >)$. That is, $P_k(p_k) \neq \perp$. On the other hand, if $P_k(p_k) = 1$, Q' should

Figure 11.2 A table of programs P_j, their corresponding binary representations p_j, and their computation results. Two special symbols ϵ and $\perp$ indicate "it terminates with no output" and "it never terminates", respectively

	p_1	p_2	p_3	p_4	p_5	p_6	...
P_1	0	0	0	0	0	0	
P_2	$\perp$	$\perp$	$\perp$	$\perp$	$\perp$	$\perp$	
P_3	0	1	0	1	0	1	
P_4	0	1	10	11	110	111	
P_5	ε	ε	ε	ε	ε	ε	
P_6	ε	$\perp$	ε	0	ε	1	
$\vdots$							

execute an infinite loop at this time because of the definition of $Q'(< p, x >)$, and $P_k(p_k) = \perp$. That is, assuming $Q' = P_k$ for some k, the behavior of $P_k(p_k)$ when p_k is given as an input to this program has no definite value.

This is a contradiction, and hence the assumption that Q exists is not true. That is, there is no program Q that solves the halting problem. □

You may first feel like being deceived. Diagonalization is a fairly elaborate proof technique and takes time to get used to. Anyway, it turns out that there are only countable infinite functions that can be computed by computer programs, and that we can define another function which is not computable by any computer program. In fact, it is known that the set of general functions considered here has the same cardinality as the set of real numbers. In other words, there are as many uncountable infinite functions as real numbers, only countable infinite functions can be computed by programs, and the halting problem is located within this difference.

11.4 Undecidability of Origami Folding Problem

Now we turn to computational origami. In the previous sections, we use the Turing machine as a computer model and the C language as a typical programming language. In such computational models, it is important that "basic operations" are clearly defined, because "a computation" is defined by a process that performs a sequence of "basic operations" in a specific order. We learned that under the restriction of the computation principle that such basic operations are repeated finite times, we can prove that countable infinite problems can be solved and there are problems that cannot be solved in principle.

We can also consider origami as a platform of computation in a sense. That is, when we fold origami, it can be thought as a kind of "computation" by folding it using some basic operations defined on paper (for example, Huzita's axioms with Hatori's operation). Then what is the computation power of a certain computational mechanism called "computation origami"? We will give a kind of answer to this question in this section. Specifically, we introduce the undecidable problem of the structure similar to the previous sections.

Since this is a simple application of diagonalization, the logic of the proof itself is not difficult if you are familiar with diagonalization, but rather simple. However, the interpretation and significance of the statement will be individual. That is, origami can be said to be capable of countable infinite types of computations as computer programs, or conversely, it can also be said that it has points which cannot be folded in principle.

. .

Aside from the interpretation or significance, let us return to the main topic. In this section, we prove that the following *folding decision problem* is a natural and simple undecidable problem in a reasonable computational origami model.

Input: An origami paper and four points p, q, r, s on it.
Output: Determine whether there are two lines ℓ_1 and ℓ_2 that satisfy the following two conditions: (1) They can be made by finite number of foldings starting from three points p, q, r. (2) They cross at point s.

Roughly speaking, the folding decision problem is only asking whether it is possible to make another point s as the intersection of the folding in finite times using the given three points p, q, r as references. This is a very natural problem on origami, but it cannot be decided. Here, we simplify this problem further and consider the following one-dimensional version of the folding decision problem.

Input: A line segment and four points p, q, r, s on it.
Output: Determine whether the point s can be creased in a finite number of foldings starting from given three points p, q, r.

Even this simplified problem is an undecidable problem. Since the two-dimensional folding decision problem completely includes the one-dimensional folding decision problem, we only deal with the one-dimensional version.

One-dimensional origami P is a line segment of finite length and zero thickness. Without loss of generality, we assume that P has length 1 and it is initially placed on the interval [0, 1]. Then any point on P is represented by a real number in the interval [0, 1]. That is, any point p on the one-dimensional origami P has some real coordinate value. This coordinate value is denoted by $P(p)$. Moreover, when unambiguous, a folding state of origami P is also written as P, and it is assumed that the left end of P is always aligned with coordinate 0. At this time, the coordinate value of p on the folded state P is also written as $P(p)$. Note that these coordinate values are real numbers, which is essential.

On two-dimensional origami, we often consider seven basic operations that are known as a combination of six axioms by Huzita and additional one by Hatori. (See [DO07, Chap. 19] for details.) In other words, an origami folding operation is the application of one of these seven axioms. As a result, one new folding line is obtained, and the intersections between this new line and the other existing lines are generated as new points. In one-dimensional origami, these basic operations are simplified as follows:

1. Specify $P(p)$ for a point p that already exists on P, and fold some paper layers around it.
2. Specify $P(p)$ for a point p that already exists on P, and open some paper layers on the other side around it.

These operations can create new points on the paper that have overlapped $P(p)$. They can be regarded as essentially the same operation. That is, it is an operation of flipping the overlapping paper on the other side around the point $P(p)$ that already exists to generate new points q. Note that $P(p) = P(q)$ holds at this time. To generate a new point, only this operation is allowed.

We consider the above operation in more detail. In the one-dimensional version of the folding decision problem, four real points p, q, r, s are given. Hereafter, without loss of generality, we assume that $P(p) = 0$, $P(q) = 1, 0 < P(r) < 1, 0 < P(s) < 1$. We call p, q, r as *start points* and s as a *goal point*. The folding operation permitted in this problem is the operation for the start points and the points generated from them. In other words, s cannot be used as the reference of each folding operation. The goal of the folding decision problem is to generate a point r' that yields $P(r') = P(s)$ over the folded state P with a finite number of operations. Note that the comparison $P(r') = P(s)$ can be done with infinite precision when a new real point r' is obtained. (If they are not equal, the result $P(r') < P(s)$ or $P(r') > P(s)$ is obtained.) That is, we assume in this model that two real points can be compared with infinite precision in a constant time. We call the real points generated from the start points *foldable points*.

First we show a theorem on the number of foldable points.

Theorem 11.4.1 *Let the coordinates of three start points p, q, r on a one-dimensional paper P be $P(p) = 0$, $P(q) = 1$, and $0 < P(r) < 1$. Then, the number of foldable points is countable.*

Proof For the set $S_0 = \{p, q, r\}$ and for each $i > 0$, we define the set S_i as follows. The set S_i includes a point t if and only if (1) t can be folded by the ith folding operation from the start points, and (2) $t \notin \cup_{0 \leq j < i} S_j$. That is, S_i is the set of points that can only be folded by i folding operations. Here, for any given constant i, there is a constant number of folded states of P after i times of constant kinds of folding operations. Thus, $|S_i|$ is also a constant number. Since the number of points included in each S_i is countable, the number of foldable points obtained by adding them all for the natural number i is also countable infinite. □

As proved in Theorem 11.2.1, the set of real numbers is not countable. Thus, Theorem 11.4.1 shows that once start points are given, there are points that cannot be folded by a finite number of folding operations. Using this fact, we can show the following undecidability.

Theorem 11.4.2 *Even if it is a one-dimensional origami model, the folding decision problem is undecidable.*

Proof To derive contradictions, we assume that there is an algorithm A that solves the folding decision problem. That is, for any input p, q, r, s, A always outputs "Yes" or "No" within a finite time. We can consider that A is written in some programming language on a computer. For the sake of simplicity, we assume that it is written in C. Once the C program $\mathcal{A}$ that implements the algorithm A is fixed, we can define a function $t_{\mathcal{A}}(p, q, r, s)$ that returns the running time of this program. That is, for any four real numbers $p, q, r, s, t_{\mathcal{A}}(p, q, r, s)$ gives the number of steps performed by $\mathcal{A}$ with input (p, q, r, s) before returning "Yes" or "No". By assumption, $t_{\mathcal{A}}(p, q, r, s)$ is finite for any four real numbers that satisfy the condition.

Here we fix the start points p, q, r as $p = 0, q = 1$, and $r = 1/\sqrt{2}$. We also define T_i by a set of real points s with $T_i = \{s \mid t_{\mathcal{A}}(p, q, r, s) = i\}$. That is, T_i is a set of real points that can be determined in i steps. We show that $|T_i|$ is countable. Note first that T_i contains two kinds of points. Of T_i, let Y_i be a set of points where $\mathcal{A}$ answers "Yes", and N_i be a set of points where $\mathcal{A}$ answers "No". According to the definition, $\mathcal{A}$ answers "Yes" in the ith step for the point s when there exist s' and $i' < i$ with $s' \in Y_{i'}$ and s' is put on s in the ith step. Thus, obviously Y_i is a countable set (and a finite set). Therefore, it is sufficient to show that N_i is also countable. If N_i contains an uncountable infinite number of points, then there is an open interval (a, b) somewhere where all internal points are included in N_i. (Otherwise, all N_i points are countable because they are isolated and numbered in ascending order.) Letting $a' = p = 0$ and $b' = q = 1$, then we have $0 = a' < a < b < b' = 1$, and a' and b' are both foldable points.

Now we fold a' on b' to obtain a new point $c(= 1/2)$. If $a < c < b$, we have a contradiction because c is an instance of "Yes". Therefore, we have either $a' < c \leq a < b < b'$ or $a' < a < b \leq c < b'$. In the former case, we replace a' by c, and in the latter case, we replace b' by c. Repeating this process, after a finite number of operations (precisely, $O(\log_2(1/(b - a)))$ foldings), we can make a new point inside the interval (a, b). However, this contradicts the assumption that all points inside the open interval (a, b) are included in N_i. Thus, N_i can contain only countable points. Thus, $|T_i|$ is countable, and hence $\cup_{0 \leq j \leq i} T_j$ is countable for any positive integer j. Therefore, we can see that the cardinality of the set of points decided by $\mathcal{A}$ in finite time is countable. Because it is a countable set, point sets decided by $\mathcal{A}$ can be ordered as $s_1 < s_2 < s_3 < \cdots$.

Now we use diagonalization to construct an undecidable real point s. Since one-dimensional origami P is an interval [0, 1], $s_1, s_2, \ldots$ can be listed as follows:

$$
\begin{aligned}
s_1 &= 0.s_{1,1}s_{1,2}s_{1,3}\ldots \\
s_2 &= 0.s_{2,1}s_{2,2}s_{2,3}\ldots \\
&\ldots \\
s_k &= 0.s_{k,1}s_{k,2}s_{k,3}\ldots \\
&\ldots,
\end{aligned}
$$

where $s_{k,\ell}$ is the ℓth decimal digit of s_k. Now we define s by $s = 0.d_1d_2d_3\ldots$ where $d_k = s_{k,k} + 1 \pmod{10}$. Then, although s is a real point on the one-dimensional origami P, it does not appear anywhere in T_i like the proof of Theorem 11.2.1. Thus, $t_{\mathcal{A}}(p, q, r, s)$ is not finite for this s. In other words, the program $\mathcal{A}$ that implements the algorithm A does not terminate for this p, q, r, s. It contradicts the assumption that A solves the folding decision problem in finite time. Therefore, the folding decision problem is undecidable even on one-dimensional origami. □

In this section, we proved that a simple decision problem is undecidable on an origami model defined very naturally. This comes from the gap between the fact that "a finite number of points can be generated by a finite number of folding operations applied on a finite number of start points" and the fact that "there are uncountable infinite number of points in continuous areas such as origami". This result gives us two different research directions.

One (practical) direction is to construct a model that allows some "error", thinking that it is not a realistic model such as infinite precision real numbers. In other words, even if the given point s cannot be folded, it is considered okay if there is a point that can be folded in the immediate region. For example, by the argument using the point c of Theorem 11.4.2 in one-dimensional origami, we can easily state an origami model that we make a point inside $[s - \epsilon, s + \epsilon]$ in finite folding operations for any given error $\epsilon > 0$ and points s. This seems to be a relatively realistic model. As a possible future subject, research of efficient algorithm for folding a point close to a given point can be considered. In this research, the goal is to reduce the number of folding operations and to reduce the extra crease lines.

The other direction is to deepen the "computation by theoretical origami model" based on the results of this chapter. For example, although it is possible to conclude that it is theoretically impossible to design a universal debugger since the halting problem of Turing machines is undecidable, it does not mean that the computation power of Turing machines is weak and useless. In fact, on Turing machine models, computational complexity theory and algorithm theory have been investigated, and various results have been obtained. Similarly, even if origami computation can only reach countable infinite points, it does not mean that the ability of origami computation models is low and weak. For example, it is known that we can solve fourth-order equations by using the seven axioms by Huzita and Hatori. This is more powerful than the computation by using a ruler and a compass, which can solve second-order equations. Little is known about the computation power when these "basic operations" are combined algorithmically. Interesting results may be obtained when considering an "origami computation model" with "basic operations" that is completely different from conventional computers.

Reference

[DO07] E.D. Demaine, J. O'Rourke, *Geometric Folding Algorithms: Linkages, Origami* (Polyhedra, Cambridge, 2007)

Chapter 12
Answers to Exercises

In this chapter, we show the answers to exercises.

Answer to Exercise 2.1.1: Let the length of an edge of a regular tetrahedron be the unit length. We first consider two types of nets by edge-unfolding of a regular tetrahedron. By Corollary 2.1.2, since both cut three edges, the total length of the cut lines is 3. Is it possible to shorten this further? If you cut the two edges at the position of twist of a regular tetrahedron, make it flat, and cut it with a perpendicular line connecting the top and bottom, the total length of the cut lines will be a little shorter because the last perpendicular line is shorter than oblique edge. Then the total length is given by $2 + \frac{\sqrt{3}}{2} = 2.866 \cdots$ (Fig. 12.1(1-2)). Considering the cut patterns of Figs. 12.1(1) and 12.1(2), we can obtain bit shorter cut lines by making the two intersections of cut lines closer like Fig. 12.1(3). Then how short will it be?

For discussion, we name the vertices as shown in Fig. 12.1(3). Let a, b, c, d be the vertices of a regular tetrahedron, p, q be the intersections of cut lines, and r be the intersection of the line segment pq and the edge ac. By symmetry, r is the midpoint of the edge ac. Here, we would like to make the tree induced by the points a, b, c, d, p, q the minimum total length of tree connecting four points $abcd$. This notion is called *Steiner tree* in the field of computational geometry. Although it is difficult to see as it is in Fig. 12.1(3), it can be understood as if it is drawn by the net as in Fig. 12.1(4) that it is a Steiner tree connecting four diamond-shaped points consisting of two regular triangles. The Steiner tree is well studied, and especially for the vertices added as points p and q (they are called *Steiner points*), it is known that all three angles are equal, that is, all angles around p and q are 120°. Conversely, if the angle is fixed, then the points p and q must move along some arcs ab and cd, respectively. Therefore, arc ab and arc cd are arcs whose circumferential angles are 120°, and points p and q are points on these arcs.

From here onward, we can proceed with elementary discussions and calculate coordinates, but the calculation itself is quite complicated. Hence, here we use a method called *Torricelli's construction* to simplify calculation. This construction is as follows.

R. Uehara, *Introduction to Computational Origami*,
https://doi.org/10.1007/978-981-15-4470-5_12

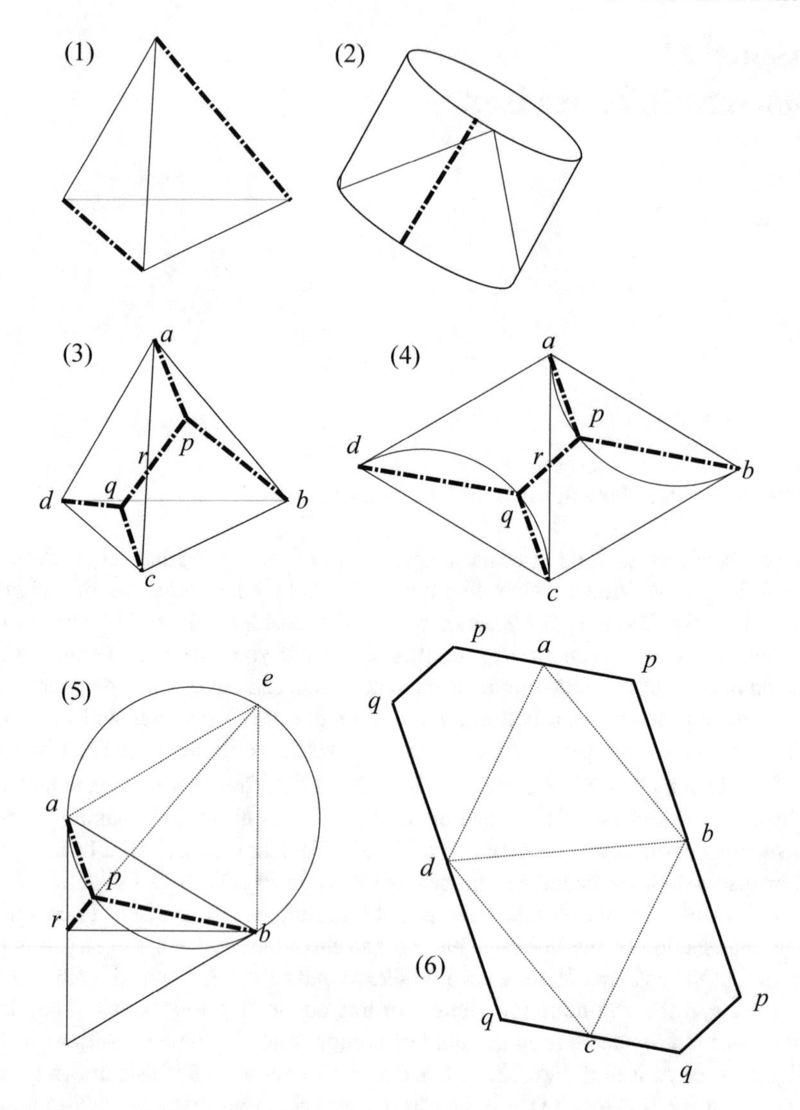

Figure 12.1 The shortest cut net of a regular tetrahedron (1)

Theorem 12.0.1 (Torricelli's construction method) *For an acute triangle $P_1P_2P_3$, put an auxiliary point X where P_1P_2X is a regular triangle. Then the point S of intersection of a circle passing through three points P_1P_2X and a line XP_3 is the Steiner point, and we have $|P_1S| + |P_2S| + |P_3S| = |P_3X|$.*

We will not go into the details of this theorem in this book. Anyway, in Fig. 12.1(4), we focus on the triangle abr, and put the point e where abe is a regular triangle. Then by drawing a circle C that passes through three points abe, we obtain the desired Steiner point p at the intersection of C and the line er. Moreover, the total cut length in this

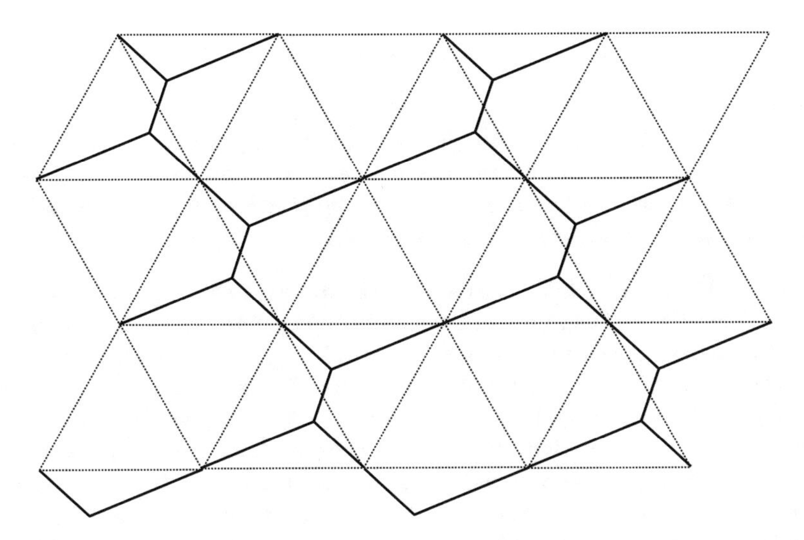

Figure 12.2 The shortest cut net of a regular tetrahedron (2)

part is $|rp| + |ap| + |bp| = |er| = \sqrt{|br|^2 + |eb|^2} = \sqrt{(\frac{\sqrt{3}}{2})^2 + 1^2} = \frac{\sqrt{7}}{2}$. Thus, the total cut length is $\sqrt{7} = 2.645\ldots$, which is the minimum value to find.

Figure 12.1(6) shows the net. Every vertex has an angle of 120°, but this optimal solution is far from a regular hexagon. The tiling by this hexagon is shown in Fig. 12.2. Theoretically, using this pattern minimizes wearing cutter when making a large number of tetrahedra.

In the case of regular tetrahedron, we can use a beautiful discussion so far. In the last step, it is possible to calculate in a straightforward manner without using Torricelli's construction, but the calculation is quite complicated in the case. In any way, the optimal solution is an impressive net. So far, for other regular polyhedra, although optimal solutions have been obtained, any smart way to these solutions is not known. The paper on the nets by the shortest cuts of five regular polyhedra is [ACN+11], and it is surprising that this result is new in 2011.

In order to find the shortest cut of a specific regular polyhedron, the basic idea of this paper [ACN+11] is simple; they first consider each possible net and construct the Steiner tree on the net, and then they find the shortest one among these Steiner trees. To put it a little further, they first list the nets, then compute the Steiner tree specifically on each of the nets, and use the one with the smallest total length among them as the optimal solution. In other words, they reduce the problem of three-dimensional polyhedra into multiple two-dimensional subproblems, then solve them separately, and compare them at the end. Of course, the correct answers are obtained; however, from an algorithmic point of view it is not a very efficient method. In the past, when I talked to Prof. Jin Akiyama, the author of this paper [ACN+11], he also seemed to think that there could be a smarter method to solve this problem. I agree with that. Considering that this problem has not been solved even for regular polyhedra until recently, it seems that general method applicable to other polyhedra has not been studied yet.

Open Problem 12.0.1 *Design an efficient algorithm for finding the net of the shortest cut for a general polyhedron. Is there any method to get the Steiner tree of vertices directly on the surface of a three-dimensional polyhedron, avoiding especially the two-step case analyses on two-dimensional polygons?*

Answer to Exercise 2.2.1: We can prove that there are only five regular polyhedra as follows. First, since all faces are congruent regular polygons, the unit regular polygon must not be more than a hexagon. In the case of hexagon, if three regular hexagons are collected at one point, it becomes a plane, and if it has more vertices, the angle of the point exceeds 360°. Therefore, the faces of a regular polyhedron are either regular triangles, squares, or pentagons. First, consider the case of regular pentagons. There must be three or more regular polygons at each vertex, but if it is four or more in the case of regular pentagons, it will exceed 360°, so only if three pentagons at a vertex can establish as a polyhedron. Thus, a regular dodecahedron is obtained. The same argument can be made about squares, and a cube can be obtained. In the case of regular triangles, we can collect three, four, and five at one vertex. In these three cases, a regular tetrahedron, a regular octahedron, and a regular icosahedron are obtained. Now all five regular polyhedra are obtained. By the arguments above, there are no other possible combinations. Thus, there are exactly five types of regular polyhedra.

Answer to Exercise 2.2.2: We first consider the dual of a cube (Fig. 12.3(1)). Place the vertices at the center of each square of the cube, and connect two vertices with an edge when the squares corresponding to the vertices share an edge of the cube. Then, the octahedron comes to be inscribed in the cube. Since the edges correspond one to one, the number of edges is 12 in both polyhedra. Also, the number six of faces of a cube corresponds to the number of vertices of a regular octahedron. Similarly, we

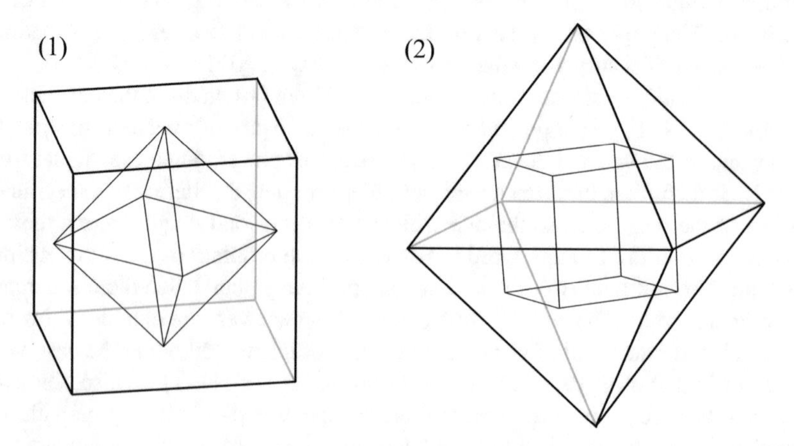

Figure 12.3 Duality between a cube and a regular octahedron

next consider the dual of a regular octahedron (Fig. 12.3(2)). Of course, the numbers of edges are again 12 each. Also, it is readily apparent that the number eight of faces of a regular octahedron corresponds to the number eight of vertices of a cube. A regular dodecahedron and a regular icosahedron are somewhat difficult to imagine, but it can be seen that they are similarly dual to each other. Also, if we do the same for a regular tetrahedron, we find that a regular tetrahedron is its own dual.

Answer to Exercise 2.3.1: At first glance, you may not find where to paste each. Here we show only the answers, so if you are not convinced, try to make a real one. First, one triangle of the tetramonohedron that can be folded from the polygon in Fig. 2.8a is an isosceles triangle, and letting that the length of one edge of the regular triangle of the original regular octahedron is 1, the length of the edge is $\sqrt{3} : \sqrt{7}/2 : \sqrt{7}/2$, which is approximately $1.73 : 1.32 : 1.32$. The lengths of the tetramonohedron of Fig. 2.8(c) are, letting that the length of one edge of the regular triangle of the original regular icosahedron is 1, $2 : 2.5 : \sqrt{21}/2$, or approximately $2 : 2.5 : 2.29$. Letting that the length of one edge of the original regular tetrahedron is 1, the size of the box in Fig. 2.8b is

$$\frac{1}{2} \times \frac{1}{2} \times \frac{2\sqrt{3}-1}{4},$$

which is approximately $0.5 : 0.5 : 0.62$.

Answer to Exercise 2.3.2: I would like you to try this once. Even if you try to aim at a regular tetrahedron by intuition, you will find that it does not go well.

Answer to Exercise 3.4.1: I would like you to make sure to actually fold the net in Fig. 3.18. In fact, two ways of folding the box of size $\sqrt{5} \times \sqrt{5} \times \sqrt{5}$ can be found surprisingly easily. On the other hand, it is pretty hard to find the ways of folding the boxes of size $1 \times 1 \times 7$ and $1 \times 3 \times 3$; they are complete puzzles. You will realize that we could not find ones without using computer.

Answer to Exercise 3.6.1: Is there any polyhedron in which the angles between two faces are not orthogonal at all, although each face is a rectangle or a square? It may be hard to imagine, but such a polyhedron exists (Fig. 12.4).

You can be convinced if you look at the figure carefully. This is a polyhedron obtained by fattened regular octahedral skeleton. It can be constructed by replacing each edge of a regular octahedron with a triangular prism and replacing the vertices with tetragonal pyramids. This polyhedron is shown in [BCD+02] in 2002. Accurate lengths of edges are difficult to calculate, but making a paper model is quite easy. First, prepare two inner faces of 12 triangular prisms. You can make 12 suitable rectangles of paper and fold them in half. Taping all of them as shown in the figure, the structure of the regular octahedron is fixed. It is easy to make the remaining outer 12 rectangles and the 6 square lids by using this frame. Pasting them together, we get one.

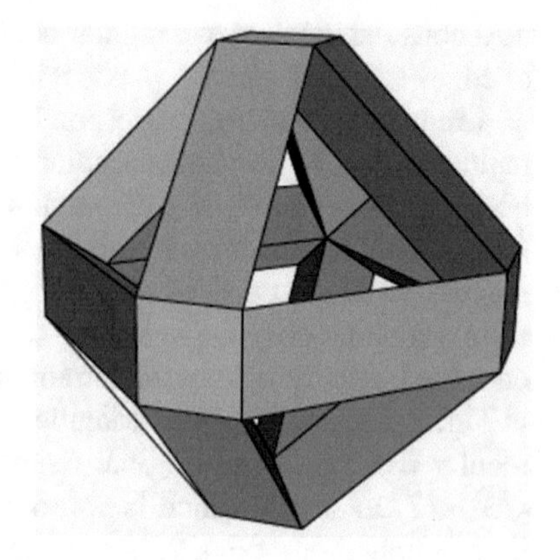

Figure 12.4 An example of a polyhedron in which the angles between the faces are not at right angles although all the faces are rectangles

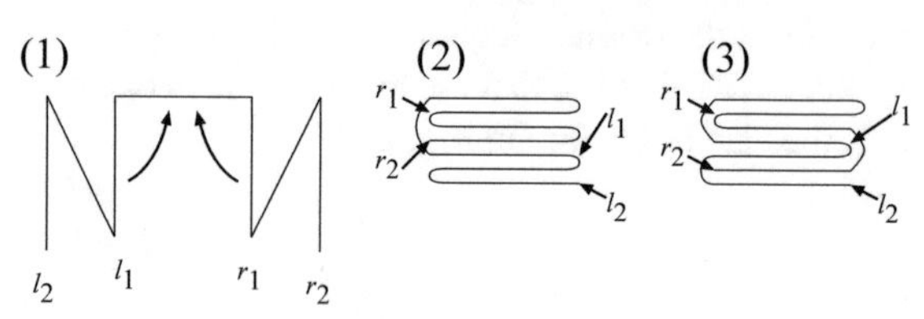

Figure 12.5 How to fold a zigzag-like pattern

Answer to Exercise 5.1.1: Consider the pattern MVM for $n = 3$ as a concrete example. Then the zigzag-like pattern $MVMMVM$ of length 6 is obtained. For this pattern, we fold the two flaps toward the center as shown in Fig. 12.5(1). Name these flaps in order from closer to the center as shown in Fig. 12.5(1). That is, the flaps appearing on the left are numbered as l_1, l_2, and those appearing on the right are numbered as r_1, r_2. After folding them, looking at the flaps from the top, we can uniquely represent how it was folded. For example, the folded state in Fig. 12.5(2) is represented by $r_1r_2l_1l_2$ from the top, and the other one in Fig. 12.5(3) is represented by $r_1l_1r_2l_2$ from the top. In this string, the order among rs and the order among ls do not get out of order; however, we can change the order between r and l as we like. This is equivalent to the number of combinations that pick two flaps from four flaps. In the current case, $\binom{4}{2} = \frac{4\cdot 3}{2\cdot 1} = 6$, and this six gives us the number of possible ways of foldings. Generalizing this argument, the answer is given by the number of combinations to select $(n+1)/2$ flaps from $(n+1)$ flaps, which is equal to $\binom{n+1}{(n+1)/2}$. With the lower bound $\binom{n}{k} > \left(\frac{n}{k}\right)^k$, it is an exponential function that can be bounded below by $\sqrt{2}^{n+1}$.

Approximations of Combinations

Although various approximations are known for the number of combinations, the following inequations are very useful: For any natural numbers $n > k > 0$,

$$\left(\frac{n}{k}\right)^k \leq \binom{n}{k} \leq \left(\frac{en}{k}\right)^k,$$

where $e = 2.718\ldots$ is the base of the natural logarithm. When we want to know a more exact value, we can use the following *Stirling's formula*:

$$n! = \sqrt{2\pi n}\left(\frac{n}{e}\right)^n.$$

Conversely, a rough approximation $n! \sim n^n$ is also useful.

Answer to Exercise 5.3.1: We first consider the pattern M^n. It is obvious when you try it, but this pattern has n distinct folded states. Trying real one is the easiest way to solve the pattern $(MV)^{n-1}MM$; you will see that there are $n + 1$ distinct folded states. These patterns seem to have the least ways of folding next to the pleat folding. Considering together with Exercise 5.1.1, these crease patterns may be "close" to the pleat folding. However, this is intuition and further study is required.

Open Problem 12.0.2 *What is a "close-to-pleat-folding" pattern with a few possible folded states?*

Answer to Exercise 6.1.1: Some people may feel strange, but the distance between the two crease lines is 1cm. If you can solve this problem easily, first guess the next question in your head and then try: Now fold the paper tape by shifting it by $a$cm as shown in Fig. 6.1a, and shift the paper tape backward by $b$cm as shown in Fig. 6.1b. Then what is the distance between the two crease lines? Does your prediction come true?

Answer to Exercise 10.1.1: Looking at the square part, you can find p2 tiling as in Fig. 12.6. Once you have determined where the center of rotational symmetry is, connect them to make a triangle. Repeating that, you will find the desired tiling. Theoretically speaking, you can fold it and make a tetramonohedron; however, it is not easy to imagine that. On the other hand, if you actually cut and fold it, you can see how easy it is.

Answer to Exercise 10.3.1: We consider folding a Latin cross P to make a non-convex doubly covered polygon. First, fold the left and right arms in half with the vertical crease lines and glue them together as shown in Fig. 12.7. Then we can construct a concave doubly covered polygon by folding the upper part and the lower part and gluing them. We can choose any real number x with $0 < x < 1$ as shown in Fig. 12.7 this time. Here x is an arbitrary real number that satisfies $0 < x < 1$,

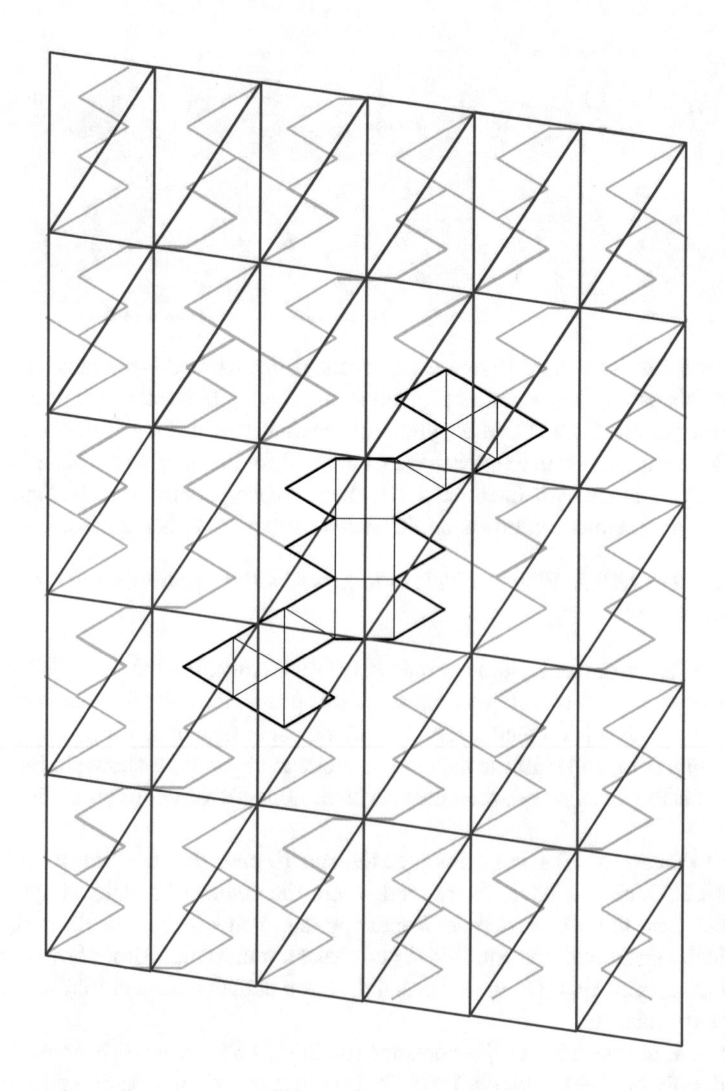

Figure 12.6 A p2 tiling made from the net by edge-unfolding of J89 and the corresponding net of the tetramonohedron

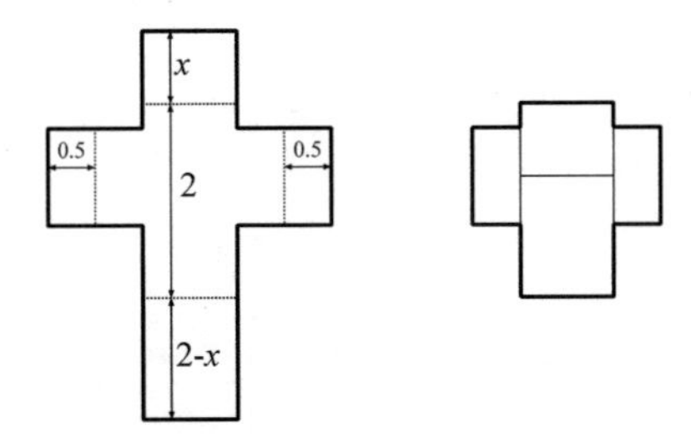

Figure 12.7 Concave doubly covered polygon that can be folded from the Latin cross. Since the length x can be arbitrary real value with $0 < x < 1$, it is uncountable and infinite

and thus there are uncountable infinite ways of folding. As far as I know, the fact that infinitely many doubly covered polygons can be folded from the Latin cross is mentioned for the first time.

References

[ACN+11] J. Akiyama, X. Chen, G. Nakamura, M-J. Ruiz, Minimum perimeter developments of the platonic solids. Thai J. Math. **9**(3), 461–487 (2011). www.math.science.cmu.ac.th/thaijournal

[BCD+02] T. Biedl, T.M. Chan, E.D. Demaine, M.L. Demaine, P. Nijjar, R. Uehara, M-W. Wang, Tighter bounds on the genus of nonorthogonal polyhedra built from rectangles, in *Proceedings of 14th Canadian Conference on Computational Geometry (CCCG 2002)* (2002), pp. 105–108

Index

R. Uehara, *Introduction to Computational Origami*,
https://doi.org/10.1007/978-981-15-4470-5

N

O

P

R

S

T

U

V

Z

Printed in the United States
by Baker & Taylor Publisher Services